THE LIVES OF *Butterflies*

THE LIVES OF *Butterflies*

MATTHEW M. DOUGLAS

THE UNIVERSITY OF MICHIGAN PRESS
Ann Arbor

Copyright © by The University of Michigan 1986
All rights reserved
Published in the United States of America by
The University of Michigan Press and simultaneously
in Rexdale, Canada, by John Wiley & Sons Canada, Limited
Manufactured in the United States of America

1989 1988 1987 1986   4 3 2 1

LIBRARY OF CONGRESS CATALOGING-IN-PUBLICATION DATA

Douglas, Matthew M., 1949–
  The lives of butterflies.

  Bibliography: p.
  Includes index.
  1. Butterflies. I. Title.
QL542.D64 1986   595.78'9   86-11284
ISBN 0-472-10078-5

Preface

The Lives of Butterflies was written in part to provide undergraduate and graduate students in entomology with a foundation and appreciation for the biology of butterflies. It was also written, in part, to serve as a general reference source for research workers engaged in various aspects of entomology, particularly the study of butterflies. It is a book that summarizes much of the important research on butterflies that has been spread out in dozens of technical journals to which most entomologists do not subscribe. It is not encyclopedic in nature, and it was never meant to be so. Rather, *The Lives of Butterflies* is intended to provide the interested reader and the professional with significant and interesting research literature.

I hope this overview, derived largely from research conducted on butterflies of the Americas, will serve as a cornerstone—a guide to what has been done and what remains to be done. My primary goal is to deliver an accurate and concise message about the lives of butterflies. In this respect, I hope that I have represented fairly the views and facts of those researchers whose work appears in this book. Any comments, criticisms, and corrections will be greatly appreciated. Any mistakes in the interpretation of facts or viewpoints are entirely my own.

In addition to the researchers who conducted the original scientific work, I would like to thank the following reviewers: Drs. Peter Atsatt, Robin Franklin Bernath, Deane Bowers, Lincoln Brower, George Byers, Frances Chew, John Downey, Paul Ehrlich, Larry Gilbert, John Grula, Jane Hayes, Joel Kingsolver, H. Fred Nijhout, Dennis Murphy, Robert Pyle, the late Dr. Robert Silberglied, Drs. Orley Taylor, Norman Tindale, Ward Watt, and Raymond White. A special thanks goes to Dr. Warren H. Wagner, Jr., and Dr.

Arthur Shapiro who read the entire manuscript at different stages for accuracy and readability. Most of their comments were incorporated into the final text. A special thanks also goes to those many researchers who answered questions by letter or phone or who supplied requests for current reprints. In addition to these researchers, a number of professional and amateur lepidopterists added greatly to the book. The late Dr. Robert Silberglied, Dr. Norman Tindale, Dr. Robin Franklin Bernath, Dr. Orley Taylor, Larry West, and Paul Douglas contributed their very fine photographs. Ms. Janee Slater aided greatly in the preparation of appendix B and the glossary. Artists Chris Baer and Susan Mason provided the black-and-white figures. To all these and others who supported me during this adventure, I offer my thanks and gratitude.

Acknowledgments

Grateful acknowledgment is made to the following publishers for permission to use copyrighted material:

Allyn Museum of Entomology for adaptations of figures from "Wing-scale Morphology and Nomenclature," by J. C. Downey and A. C. Allyn in the *Bulletin of the Allyn Museum,* no. 31, 1975.

Chanticleer Press for artwork from *Butterflies,* by T. C. Emmel, 1975.

Doubleday & Company for nine illustrations from *The Butterflies of North America,* by W. H. Howe. Copyright © 1975 by Doubleday & Company, Inc. Reproduced by permission of the publisher.

W. H. Freeman and Company for artwork from "The Color Patterns of Butterflies and Moths," by H. F. Nijhout in *Scientific American.* Copyright © 1981 by Scientific American, Inc. All rights reserved.

McGraw-Hill Book Company for adaptation of artwork from *Principles of Insect Morphology,* by R. E. Snodgrass, 1935.

Psyche for adaptations of artwork from "External Sex Brand Morphology of Three Sulphur Butterflies (Lepidoptera: Pieridae)," by R. S. Vetter and R. L. Rutowski, 1978.

William C. Brown for artwork from *How to Know the Butterflies,* by P. R. Ehrlich and A. H. Ehrlich, 1961.

Acknowledgment is also made to the following authors for permission to use their copyrighted artwork: W. Calvert, S. Dalton, J. W. Grula and O. R. Taylor, J. G. Kingsolver and T. L. Daniel, N. B. Tindale, L. T. Wasserthal.

Introduction

Through all my years of fascination with butterflies, I have always wondered how and why insects developed wings. Equally important perhaps, why do butterflies in particular possess such expansive and brilliantly patterned wings? When I was a graduate student at the University of Kansas, I quickly discovered that many entomologists before me had puzzled over these same questions.

The more I thought about it, the more I realized that there are really several different aspects to each question. For example, from which structures on the primitively wingless insects did the first winglike expansions develop? And what were the selective pressures of nature that sequentially modified these winglike expansions, or *winglets*, into functional wings as varied and beautiful as those of butterflies? Such questions have launched entomologists into arguments of awesome complexity.

After over a hundred years of disagreement, most entomologists have settled on two very general theories that explain the origin of insect wings. The majority clings to the *paranotal lobe theory*. This theory proposes that the wings of primitively wingless insects first arose as flat, broadly attached flaps that projected outward from the sides of the thorax. The *thorax*, composed of amalgamated segments that now house the leg and wing muscles, lies between the head and the abdomen. The primary function of the thorax in modern insects, therefore, is to provide locomotion—whether it be walking, running, swimming, or flying.

The paranotal lobe theory seems to make sense, especially when one considers that insects with small thoracic lobes or winglets are well represented in the fossil record. In fact, some extinct immature forms of insects had three pairs, rather than two pairs, of these broadly attached winglets. Many entomologists assume that these winglets were without joints—nonarticulated—and therefore immobile.

However, several researchers, led by Kukalovà-Peck (1978), have painstakingly reexamined the winglets of many fossilized immature insects. She has concluded that the evolving winglets were always articulated at the base and therefore movable. She further claims that the apparent nonarticulated state of developing wings common to insects such as grasshoppers (which undergo a gradual change in shape during metamorphosis) is a more recent evolutionary development that took place long after ancestral insects had developed true flight. To Kukalovà-Peck, developing articulations in immobile gliding winglets is about as logical as suggesting that eagles developed joints in their wings by bending their wing bones in flight.

There is no consensus as to how the winglike extensions first evolved. Convincing fossil evidence simply has not been discovered yet. However, most entomologists would probably agree that some sort of lateral growth—whether articulated or not—gave rise to the winglets of primitive insects. One or possibly two of these ancient groups of insects equipped with winglets ultimately became the common ancestor of all contemporary winged insects. But once winglets evolved, what functions might they have served, and more importantly, what selective pressures might have transformed these lobelike winglets into truly functional wings?

Here, as you might suspect, entomologists have had another intellectual field day. But once again, the necessary intermediate stages required to answer the questions are lacking in the fossil record. The number of explanations seems to be inversely proportional to the amount of solid physical evidence. Since virtually no evidence is available, an almost infinite number of hypotheses can be entertained. For example, several researchers have suggested that movable winglets served to protect the *spiracles*, or openings, that lead to the air supply tubules, or *tracheae* (see chap. 3).

Others claim that the evolution of any thoracic flange might have aided aerial dispersal by minute insects (Wigglesworth 1963). Still others contend that winglets may have been used during sexual displays to lure mates, as is indeed the case with some contemporary insects (Alexander and Brown 1963). Each hypothesis has its merits, but none can explain why minute winglets continued to enlarge to a size where gliding or flapping flight could be attained (Douglas 1981).

In 1974, some simple experiments suggested to me that these winglets may have enlarged in certain groups because they were useful in elevating body temperature. The work of many researchers before me had already established that the activities of insects are largely dependent on the external or ambient air temperature and the amount of solar radiation available. Because of their minute mass, the tiniest insects never attain a body temperature higher than a fraction of a degree above ambient. Their rate of heat gain is rapid, but because they have a very large surface-area-to-volume ratio, they also cool rapidly. By contrast, some large insects, particularly those that are well insulated such as bumblebees and sphinx moths, can maintain a body temperature between 32 and 38°C (90 to 100°F) even under freezing conditions (e.g., Heinrich 1979).

But the fossil record indicates that those immature insects equipped with three pairs of winglets were neither extremely small nor extremely large. In fact, their body shape and length were similar to those of modern silverfish—an inch-long cylinder whose abdomen terminates in a long central filament flanked by two long segmented tails, the *anal cerci*.

Now it just happened that I was exploring the temperature-regulating (thermoregulatory) strategies, of butterflies. One of the species I worked with was the ubiquitous Orange Sulphur butterfly (*Colias eurytheme*). This species has physical features similar to those of the fossilized insects with primitive winglets. Of course, the wings of *eurytheme* butterflies are large and obviously functional, while the winglets of the extinct insects were tiny and probably could not support the animal in either gliding or flapping flight. However, the winglets of the fossilized insects were thickened at the base and broadly attached to the thorax (Kukalovà-Peck 1978) as are butterfly wings.

If the thermoregulatory hypothesis is correct, even small winglets should increase the absorption of solar radiation, thereby heating the body more rapidly than one without winglets. Actually, some extinct species had winglets, or structures physically similar to winglets, on both sides of the thorax and abdomen. But only those on the thorax could be especially beneficial since rapid heating of the thoracic region would allow the leg muscles to warm rapidly and maintain a higher temperature. A higher thoracic temperature under cool conditions thus would enhance the efficiency of the leg muscles.

Efficient locomotion in wingless insects—including those forms that had winglets but could not fly—might have been critical to survival. A cold, sluggish insect discovered by a predator will soon be a dead one. Insects that can quickly elevate their body temperature under cool conditions can become active earlier and remain active longer. As long as the sun is shining, they have an advantage over wingless forms. Increased activity allows more time to find food and to locate mates. Likewise, higher body temperature quickens reflex action vital to predator escape (Douglas 1981).

Wasserthal (1975) and I (Douglas 1978) had previously determined that only the *basal area* of a butterfly's wing—the thickened area attached to the thorax—could be used to elevate thoracic temperature. These basal areas often bear a thick pile of black scales and hairs that enhance the light-absorbing properties of the wings (e.g., Watt 1968, 1969). I found that butterflies with intact wings achieved a thoracic temperature 30 to 50 percent higher than specimens of the same size without wings under similar thermal conditions (Douglas 1978).

Further analysis of how butterfly wings actually increased body temperature suggested that their thermoregulatory capacity is purely a physical phenomenon. That is, the sluggish flow of insect blood (*haemolymph*) through the wing "veins" (see chap. 3) has nothing to do with energy transfer between the wings and the thorax. A butterfly's wings absorb radiant energy from the sun and surrounding matter and transfer some 20 percent of this energy directly to the body by physical contact with the basal areas. More importantly, however, the expansive wings trap warm pockets of air and thus buffer the thorax from drafts. The wings also simultaneously reduce heat loss by increasing the mass of the thorax, and by modifying the flow of cooler air around it (Douglas 1978).

The experiments with *eurytheme* suggested that the winglets of the extinct ancestral insects might have increased in size because they were important thermoregulatory structures, at least in terrestrial, diurnally active insects. But in order to prove that winglets could be used for thermoregulation, a direct comparison must be made of the thermoregulatory abilities of an insect with winglets to one that does not have winglets. The test insects must have the same physical characteristics except for the presence or absence of wings, and they must be tested under identical thermal conditions.

How could such requirements be satisfied? With *eurytheme* butterflies, of course! I temporarily dropped my field work and planned the experiment. I first froze *eurytheme* butterflies and then snapped their wings off neatly at the base. A minute hypodermic thermocouple (a needlelike thermometer) was then inserted into the center of the thoracic muscles from below. Next, I cut out rectangular 3-×-5 millimeter (0.12-×-0.20-inch) sections of the thickest part of *eurytheme* wings, and impaled these wing fragments on fine insect pins. These artificial winglets corresponded to the thickened, dark winglets found on the fossil insects. The pins holding the winglets were held by two aluminum clips mounted on a balsa wood platform so that the clips could be rotated away from the body, leaving it suspended between the impaled wing fragments.

In a temperature-controlled room, a lamp that approximated the spectral qualities of sunlight was adjusted until it delivered energy equivalent to that of the noonday sun in the temperate region. Because the wings could be rotated away from the body or pressed to the sides of the thorax, I could determine how much energy the tiny winglets delivered to the thorax. This was accomplished by comparing the final thoracic temperature reached by the same butterfly body with and without artificial winglets under identical orientations and thermal conditions.

The results are striking. Without the 3-×-5-millimeter thoracic winglets, a horizontal *eurytheme* body reached a temperature of only 13.3°C above ambient after three minutes. But when the thoracic winglets were rotated and pressed against the thorax, a final body temperature of 21.1°C above ambient was attained in the same orientation, even if contact between the winglets and thorax was not perfect. Another series of experiments established that increasing the size of the winglets increased the temperature excess of the thorax, but in a manner similar to the economic law of diminishing returns.

Thus, there is a physical size limit beyond which little additional heating advantage is obtained by any further increase in winglet length. For *Colias eurytheme* butterflies, this size limit is about 10 millimeters (0.4 inch) in wing length for a body 15 millimeters (0.6 inch). Wings of greater size confer little additional thermoregulatory advantage. Because *eurytheme*'s wings are typically 30 millimeters (1.2 inches) in length, this means that only the inner or basal one-third of the wings can be modified through evolution for thermoregulatory purposes (Douglas 1981).

So the wings of *Colias eurytheme* butterflies may have solved a nagging mystery about the selective pressures that led to the original expansion of tiny lateral winglets on primitive insects to a size at which gliding or flapping flight was possible. The thermoregulatory hypothesis also explains why winglets located along the sides of the abdomen did not expand and were ultimately lost in modern insects: There were no leg muscles in the abdomen of primitive insects and therefore little need for additional thermoregulatory structures there. In the thorax, by contrast, the muscles required to move the winglets were evolving concurrently with the thoracic leg muscles, and any lateral, broadly attached lobes would have been beneficial here.

In fact, the fossil record shows that the winglets of virtually all ancestral insects were very broadly attached, dark, or obliquely striped with the darkest portion lying adjacent to the thorax. The experiments with artificial thoracic winglets made from *eurytheme* wings show that the thoracic winglets of extinct insects would have conferred a similar thermoregulatory advantage on a terrestrial, diurnally active insect. This makes good sense physiologically too, because the metabolic furnaces of contemporary (and primitively) wingless insects—such as the common silverfish—cannot generate enough excess energy to warm the body. Yet, nearly all these insects depend on warm environmental conditions for optimal activity.

It is more than probable, then, that terrestrial insects of four hundred million years ago were elevating their body temperature by basking as insects do today. Basking is a reliable and energetically inexpensive means of thermoregulation. By fine-tuning their orientation to sunlight, by picking protected places away from chilling winds, and by remaining close to the basking site—logs, rocks, or even large leaves—ancient insects could trap pockets of warming air beneath their thoracic winglets while simultaneously protecting the warm thorax from wind-mediated heat loss.

The ability to thermoregulate efficiently would have had many selective advantages. The winglets would have increased an insect's thoracic mass and thus its ability to gain and retain heat. Metabolically, high thoracic temperatures cost it nothing. And what a tremendous payoff! Higher thoracic temperatures translate into greater activity, in turn permitting longer foraging periods, quicker escape from predators, and increased likelihood of locating and competing successfully for mates.

Indeed, greater individual activity would have translated into greater dispersal opportunities for the entire species. This is a selective advantage at the species level that ultimately would be a tremendous advantage to all their descendants. Given certain environmental conditions and basking behaviors, the conclusion is inescapable that any broadly attached thoracic winglets would have aided in heating the thoracic leg muscles of primitively wingless insects (Douglas 1978).

Of course, the selective advantage for thermoregulation does not exclude other possible winglet functions that may have evolved concurrently as more expansive thoracic winglets were selected for thermoregulation. If thoracic winglets expanded in response to a thermoregulatory pressure, a size would eventually be reached at which the developing wings could be used in conjunction with the long anal cerci and central filament of the abdomen to perform efficient gliding maneuvers from the tall vegetation of the Devonian (see app. A). This would have led to the structural state necessary for the evolution of truly functional wings (Douglas 1981). A recent series of experiments by Kingsolver and Koehl (1985) supports this scenario.

That winged insects were a consummate success is illustrated by the fact that well over 95 percent of the estimated one million living species of insects have wings as adults or, if wingless, evolved from ancestors that had wings. The primitively wingless species of today are restricted largely to a groveling existence, rummaging through soil and leaf litter, while their winged relatives virtually rule the world in terms of abundance and diversity of species.

Contents

CHAPTER I

From Stoneflies to Butterflies

The evolution of insect wings led to an explosion of diversity in insect form. Wings allowed insects to disperse at will, to find patchy resources for themselves or their offspring, and to utilize a myriad of different terrestrial environments. Some entomologists contend that the evolution of insect wings was a singular event—that all contemporary winged insects descended from a single species, even though many different but related groups of insects with thoracic and abdominal winglets likely existed. Others contend that the dragonflies (Order Odonata) evolved independently of all the other winged insect groups, basing their decision on differences in wing articulation, flight mechanism, wing *venation*, and other characteristics.

However, there does seem to be a general plan of wing venation (except in highly modified forms) present in all extinct and extant insects. The wings' *veins* are really nothing more than extensions of the tracheal air tubes that ramify from the external spiracular openings to all parts of the insect body. Not only is the general sequence of veins from the front to the back of the wings more or less the same in all orders of insects, but certain veins are elevated with respect to others so that the arrangement resembles the ridges of corrugated cardboard when viewed from the side. It is very unlikely that the same sequence of wing veins and three-dimensional pattern would result if wings evolved independently more than once (Byers, personal communication).

Yet there is another puzzle. Extinct insects thought to be ancestors of modern winged forms had three pairs of thoracic winglets, one pair for each thoracic segment. Modern insects, however, have only two pairs, one each on the *mesothorax* and *metathorax*, respectively the second and third thoracic segments. What happened to the winglets on the *prothorax*, the first thoracic segment?

A common argument is that flight muscles could not develop in the prothorax because this would require that the head be fused with the prothorax. The reason for this is that the flight muscles in the more advanced groups of insects, including butterflies, do not attach directly to the wings. Instead, *indirect muscles* alternately depress the top of the thorax, pushing down on the wing bases to raise the wings, and then shorten the thorax, thereby humping the top and raising the wing bases to lower the wings. Thus, the wings are analogous to levers while a thoracic process serves as a fulcrum. But the change in the shape of the thoracic capsule is accomplished by these two sets of indirect, *antagonistic muscles.*

The *longitudinal muscles* running lengthwise within the thoracic segments require two points of attachment—something solid against which they can pull. The attachment sites are provided by broad, platelike *phragmata* (*phragma* = singular) that invaginate between the metathorax and mesothorax, and the mesothorax and prothorax. However, the longitudinal muscles between the prothorax and the head have remained small because there apparently never was an anterior phragma or other attachment site. As a result, the prothoracic winglets—which may have originally protected the air-breathing gills or the prothoracic spiracular openings of aquatic or semiaquatic insects—became secondarily immobile and never developed into wings. Over the course of millions of years, the prothoracic spiracles apparently disappeared.

No one really knows if this was in fact the reason for wings developing on the mesothorax and metathorax but not the prothorax. Intuitively, however, it makes some sense. If wings had developed on the prothorax, they would require the necessary muscles to move them, and that would necessitate a stationary phragma or other point of attachment on the back of the head. This development would have severely limited the movement of the head, itself a structure greatly modified for feeding and sensing the environment. These basic functions are of critical importance to survival, and the organs that provide these functions cannot be easily altered.

There are other smaller muscles that control the attitude of the wings during flight. These muscles are attached directly to the wing base and can move the wings forward and backward, as well as tilt them while the insect is flying. In the most primitive groups of flying insects, such as the dragonflies, the forewings and hind wings are moved independently. However, most of the advanced orders with two pair of wings (such as the butterflies) have developed some sort of coupling mechanism so that the wings are moved as a unit.

The butterflies have no physical coupling mechanism, but their wings are so broad that they overlap considerably and thus function as a unit during flight. However, one of the Australian "skipper" butterflies (*Euschemon raffesia*) has a perfectly formed *frenate* wing-coupling mechanism, in which the top of the hind wing has a lobe that fits into a slip of modified scales on the bottom of the forewing. This frenate coupling mechanism is similar to those found in many families of moths, including the butterflylike moth family Castniidae.

The indirect wing muscles of butterflies respond with one contraction for each nervous impulse that reaches them. Such *synchronous* (one nervous impulse = one muscle contraction) *muscles* limit the number of wing beats that can be produced each second. In the hummingbirdlike sphinx moths (family Sphingidae) the synchronous muscles can beat the wings about eighty to one hundred times per second (Heinrich 1971a, 1971b). This is fast enough to make the wings appear blurred as the insects hover over deep-throated flowers in search of nectar.

But such feats pale before those of some minute flies called midges (family Chironomidae) that can beat their wings nearly one thousand times per second. The reason for such a rapid wing beat rate is that the muscles contract *asynchronously*, that is, for each nervous impulse the muscles contract and relax many times. In effect, the rapid contraction and relaxation of the indirect flight muscles causes the thoracic capsule to resonate—so that the natural frequency of the thoracic capsule becomes the same as the frequency of muscle contraction/relaxation. In fact, the wing beat rate of an insect can be measured very accurately by determining the pitch of the sound produced during flight.

How do butterflies compare to these whirling dervishes of the insect world? By comparison with those of true flies, the wings of most butterflies are very wide and relatively broadly attached to the thorax. In addition, butterfly wings are operated primarily by the synchronous indirect flight muscles. As a group, therefore, butterflies average only five to ten wingbeats per second. Faster wing-beats are possible in some species, but there is a limit beyond which the edges of the wings would be torn to shreds. Wingbeat frequencies typically increase in older, tattered butterflies because tattered wings cannot generate the same amount of lift and thrust as fresh, new wings.

In general, those insects with narrowly attached and slender wings are capable of much greater speeds and hovering abilities than butterflies. But there are few insects that can outdo the larger butterflies at gliding. Just witness the migrating Monarchs (*Danaus plexippus*) this fall riding the thermals over fields and buildings, effortlessly gliding hundreds of yards at a time, before another contraction/relaxation of the flight muscles is necessary.

All true butterflies possess wings, but from which group of winged insects did the butterflies arise? This is a difficult question to answer because like all delicate, invertebrate animals, butterflies are rarely fossilized, and those fossils that have been found are usually not in the best of condition. As a result, our knowledge of the evolutionary lineage of butterflies is very incomplete.

We do know that winged-insect fossils appear in carboniferous (see app. A) deposits over 50 million years before birdlike reptiles ever attempted flight. And by this time, there was already a diverse assemblage of winged insects, most orders of which are now extinct. But the Lepidoptera were not among these ancient winged insects. Thus, we must infer what happened between the appearance of the first winged insects well over 350 million years ago and the appearance of the butterfly-moth lineage perhaps 175 million years ago.

By observing the similarities in wing venation and other physical characteristics between the Lepidoptera and other extant orders of insects, we can arrive at a reasonable inference as to when and from which groups the Lepidoptera descended. Although many contend that primitive mothlike insects arose from a caddisflylike insect, others suggest that a stoneflylike ancestor gave rise to both caddisflies (Order Trichoptera) and the Lepidoptera. These developments took place sometime between the end of the Permian and the Jurassic periods (see app. A). During this time period, the land was invaded and colonized by truly terrestrial plants.

One winged-insect fossil, *Eoses triassica* Tindale, was found in

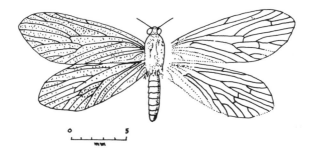

1. *Eoses triassica* Tindale, an early mothlike insect from the Upper Triassic. This restoration is based on a specimen from the Mount Crosby insect shales of Australia: the wing bases and body are hypothetical. (From Tindale 1981.)

upper-Triassic rock beds and deserves some consideration. The wings of *E. triassica* resemble those of *Mesochorista proavita* Tillyard, a more recent fossil stoneflylike insect, also from the Upper Triassic (see app. A). Yet, despite the similarities in wing venation, there are considerable differences, especially in the number of small *cross veins* in the wings of *E. triassica*. The cross veins bridge the longer veins running from the bases to the margins of the wings. The three known wings of *E. triassica*, discovered in Queensland, Australia, are sufficient to give us a useful picture of the early lepidopteran stem, which must have diverged from the stonefly line (Order Plecoptera) about 215 million years ago (Tindale 1981).

These fossils indicate that the lepidopteran line arose during the Triassic at a time when the ancient southern "supercontinent" Gondwana—comprising South America, Africa, India, Antarctica, and Australia—was still intact. Tindale has also recently described several other fossil insect wings that likely belonged to early lepidopterans. These fossils reinforce the hypothesis that both the caddisflies and the butterfly-moth line emerged as parallel lineages from a stonefly ancestry. Flowering plants also had their origins in the Triassic, and because the vast majority of extant Lepidoptera are so intimately associated with flowering plants (e.g., Ehrlich and Raven 1965, 1967), it is reasonable to assume that both groups evolved together. This would make the appearance of primitive Lepidoptera contemporary with that of the therapsids, a group of reptilelike mammals that ultimately gave rise to true mammals. So we and the butterfly-moth line apparently share a common evolutionary time scale.

The origins of true butterflies—the *Rhopalocera* or *Papilionoidea*, depending upon the system of classification—are nearly as obscure as the origins of the stoneflylike insects that gave rise to the primitive moths. Thus, it must be understood that a great deal of outright speculation takes the place of hard fact when it comes to discussing their evolutionary origins. The first true butterflies are dated from Tertiary deposits (see app. A) over fifty million years old. Many of these fossils can be assigned to contemporary families of butterflies. No one knows for certain how the wings of these primitive butterflies were colored, but judging from the wings of primitive stoneflies, the first true butterflies were patterned in whites, blacks, and shades of brown (Durden and Rose 1978; Tindale 1980).

On the basis of early tracheal patterns found in developing butterfly *chrysalids* or *pupae*, some investigators propose that the swallowtails (family Papilionidae) are the most primitive living group of butterflies. In fact, Durden and Rose (1978) have discovered Eocene (see app. A) fossils of Papilionidae in western North America that closely resemble *Baronia*, a primitive genus of swallowtails whose single described species resides in central Mexico, south of Mexico City. If this is so, then the *Baronia*-like swallowtails likely evolved nearly fifty million years ago.

The Mexican species, *Baronia brevicornis*, is truly an enigma. Most of its characteristics are in line with those of true swallowtails, but without the very short hind wing tails it would superficially resemble an ancient subfamily of swallowtail butterflies, the Parnassiinae, which typically inhabit arctic or alpine regions. Yet, closer examination reveals that *B. brevicornis* has certain characteristics of wing venation found in pierids (the whites) and the nymphalids (e.g., the anglewings). It is thus apparent that *Baronia*-like butterflies diverged from the ancestral swallowtail group before the true swallowtails arose. Like the parnassids, the *Baronia* larvae pupate under rubble or actually tunnel into the ground—a bizarre behavior for a swallowtail (Tyler 1975).

B. brevicornis, then, is a "living fossil" that occupies a biological status similar to that of the *coelacanth*, the primitive lobe-finned fish—whose descendants gave rise to lungfish and the first amphibianlike animals—thought to be extinct for three hundred million

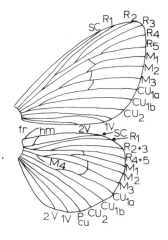

2. Hypothetical wing venation for the archetypal butterfly. (From Tindale 1980.)

years. Biological anachronisms like the coelacanth and *B. brevicornis* are not "missing links" in the evolutionary record. They represent ancient lineages of animals whose members have largely died out. They are the only surviving members of formerly diverse groups of organisms.

The fact that the earliest butterfly fossils are similar to extant *Baronia* provides a clue to the origin of true butterflies, as a loosely defined taxonomic group to be sure. Because mothlike insects appear in the fossil record over a hundred million years before butterflies, and because moths share so many uniquely derived traits with butterflies (such as broad wing scales and a tubular *proboscis* derived from the *galeae* of the *maxillae*—see chap. 3), it is reasonable to assume that the *Baronia*-like butterflies arose from a day-flying group of moths with similar wing venation and antennae form. Indeed, a hypothetical archetype of the ancestral butterfly's wing venation, based on patterns of tracheation (veins) in different families of butterflies, bears a striking resemblance to that of *Castnia licoides*, a member of the butterfly-moth family, Castniidae. Contemporary castniids are tropical day-flying moths that bear an uncanny resemblance to butterflies (Tindale, personal communication).

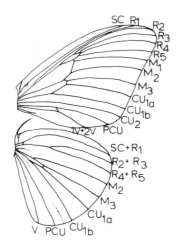

3. Wing venation of a male *Castnia licoides* Boisduval. Compare this diagram with the hypothetical wing venation for the archetypal butterfly wing in figure 2. (From Tindale 1980.)

While we will probably never know precisely which ancestral forms of stoneflies gave rise to the moths, and which primitive day-flying moths gave rise to the true butterflies, one fact is certain. Over the known fifty million years of their history the true butterflies have evolved into the most striking group of aerial insects that have ever graced this earth. The beauty of form and diversity of wing patterns and colors found among the twenty thousand or so described species are unparalleled in all the animal kingdom.

CHAPTER 2

The Immature Stages

A butterfly's *metamorphosis* is nothing like the bizarre transformation experienced by Gregor Samsa in Kafka's short story *The Metamorphosis*. It is far more spectacular and complex. The precision developments that take place within each immature stage—egg, larva, and *chrysalis* or pupa—are regulated by different sets of genes in symphonic coordination. Yet, the genetic constitution of cells within the butterfly egg is the same as that within the cells of each subsequent stage. During metamorphosis then, some genes must be turned on while others are turned off, and it is this regulation of gene expression that orchestrates the different stages of metamorphosis.

As with all life cycles, there is no "beginning" in a butterfly's metamorphosis. After all, a cycle is a cycle, and who is to say that the butterfly is not the egg's way of making another egg? Certainly it makes some evolutionary sense to view it that way. Tradition, however, dictates that discussions of metamorphosis and the life stages begin with the egg. I will not buck tradition.

THE EGG

A butterfly's life cycle—like that of most sexually reproducing organisms—begins as an egg and sperm, each of which contains exactly one-half the number of chromosomes (haploid) found in the body cells. The minute eggs are produced within several tubular ovaries that lead to lateral oviducts. The oviducts then meet and fuse to form a single duct leading to the outside. Just prior to oviposition, the egg is fertilized by a sperm cell that has been stored in the female's internal receptacle called the *corpus bursae* (see also chap.

4. A smooth spherical egg of the Old World Swallowtail (*Papilio machaon*).

3). Sperm travels from this storage chamber via a special tube leading to the egg in the lower part of the oviduct.

Eggs are typically laid on or near an appropriate larval food plant whose suitability is determined through a combination of visual, olfactory, and tactile cues. In addition, egg placement on the food plant is often characteristic for a given species. For example, females of the Spring Azure butterfly (*Celastrina argiolus pseudargiolus*) lay eggs one at a time among the buds and opening flowers of many plant species. By contrast, females of the Great Spangled Fritillary (*Speyeria cybele*) often flop to the ground in open shaded areas where violets blanket the ground. Here they lay eggs on nearby sticks, dead leaves, and even logs, but seldom directly on the food plant. The newly hatched larvae instinctively search for and locate tender young violet leaves.

Butterfly eggs are usually white or off-white, but some butterflies

5. The highly sculptured, dome-shaped egg of the Painted Lady butterfly (*Vanessa cardui*).

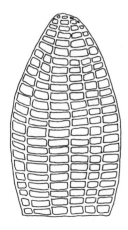

6. The ridged eggs of the Falcate Orangetip butterfly (*Anthocharis midea*) are well camouflaged on the leaves and flowers of their various cruciferous host plants.

such as the Red Admiral (*Vanessa atalanta*) have green eggs, while those of the Common Sulphur (*Colias philodice*) produce yellowish eggs. Furthermore, many eggs change color with age, often noticeably just prior to hatching. Also, eggs parasitized by the larvae of tiny wasps and flies may be darker than nonparasitized eggs. Finally, there is some evidence that egg color may be determined by genes that affect the color of the egg's cytoplasm, or the color of the maternal coats overlying the "eggshell" (*chorionic* coats). However, variations in color may also be produced by compounds secreted by special accessory (*colleterial*) glands of the female. These glands secrete an adhesive cement around the egg during oviposition to ensure that the egg sticks to the surface upon which it is laid.

In addition to color variations, butterfly eggs come in a number of shapes ranging from spherical to conical, barrel-shaped to turban-shaped. In some species, the eggs are camouflaged with long anal scales and hairs from the female's abdomen (Nakamura 1976). And although many butterfly eggs are relatively smooth, some lycaenid and riodinid eggs are masterpieces of architectural design. Downey and Allyn (1981) have examined the fine structure of the outer chorion surface—the "eggshell"—of lycaenid and riodinid eggs with

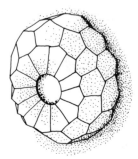

7. The sculptured egg of Rawson's Metalmark (*Calephelis rawsoni*) shows an intricate surface design, the product of the female's ovarian follicular cells. (From Downey and Allyn 1981.)

a scanning electron microscope. These eggs are highly sculptured, and although there is much variation between species, eventually it might be possible to base identifications on the three-dimensional patterns of the chorion surface.

Of course, the chorion pattern is really an expression of an adult characteristic, because its architecture is a product of the ovarian follicular cells within the female. Jagged ridges, crests, and spiny protuberances are the rule rather than the exception in both groups. A representative sample is shown in figure 7. Notice the regularity of the sculptured pattern, and the central depression with a tiny pore or *micropyle,* where the sperm can penetrate the chorion. The micropyle is often surrounded by beautiful geometric designs, commonly a flower-shaped pattern with five to seven petals (Downey and Allyn 1981).

The egg chorion consists of several layers and is composed of a number of proteinaceous compounds that make it resistant to puncture and abrasion, while allowing the exchange of oxygen and carbon dioxide for the rapidly developing embryo. This exchange of gases can be accomplished by simple diffusion through ducts in the chorion layers, or by means of a bubble of air or plastron that forms a regular porous grid. The plastron acts like a physical gill, allowing the diffusion of gases even when covered with water, and also helps prevent desiccation. In those eggs that lack plastrons and special *aeropyles* or breathing pores, respiration is accomplished in part through the micropyle (Wigglesworth 1972).

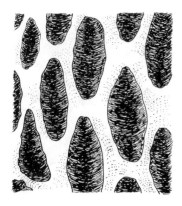

8. A close-up of the eggshell plastron in the Eastern Pigmy Blue (*Brephidium pseudofea*). Note the fine, symmetrical detail of the pores and the ridges comprising the plastron. (From Downey and Allyn 1981.)

A *vitelline membrane* lies beneath the chorion and surrounds the developing larva. This membrane is derived from the cell wall of the ovum and also helps to prevent desiccation. Other protective layers develop between the membrane and chorion as the embryo within forms distinctive germ tissue layers from which specific organs and organ systems will arise. As development progresses, the original color of the egg darkens until the embryo may be completely outlined just prior to hatching.

The transformation from a single fertilized cell to a minute larva containing thousands of cells takes anywhere from three to ten days under optimal summer conditions. However, some species lay eggs in the fall that *diapause*—that is, they remain in a state of arrested development until spring. Low humidity, low temperatures, and shorter day lengths may alone or in combination arrest or slow larval development within the egg.

Up until hatching time, the larval cells multiply at a geometric rate, utilizing the food stored within the yolk. When development is complete, the larva uses its strong toothed "jaws" or *mandibles* to gnaw its way through the egg membranes and the chorion, sometimes along a predetermined line of thinness or weakness near the top of the egg. The exit flap may be completely eaten, or a circular chewing pattern may produce a "flip top" cap.

After emerging, the larva may consume the remainder of the

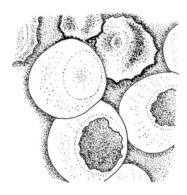

9. A magnification of the eggs of *Euselasia hieronymni* showing eclosion holes and eggs that have been devoured down to their bases. (From Downey and Allyn 1981.)

chorion, and even the chorion of adjacent undeveloped or unhatched eggs as its first meal. Considering that female butterflies may lay between one hundred and two thousand eggs, depending on the species, cannibalism seems to be of little consequence. In larger species, however, natural selection may have favored single egg oviposition as a response to cannibalism. Generally, single egg oviposition ensures an adequate food supply for a larva stranded on a small food plant.

THE LARVA

Newly hatched larvae that do not diapause begin to consume their host food plant as if each day were their last. However, not all butterfly larvae are vegetarians. For example, the caterpillars of the North American Harvester butterfly (*Feniseca tarquinius*) are entirely carnivorous on wooly aphids and related insects. Still other exotic lycaenids dine on ant larvae and pupae in a peculiar symbiosis (see chap. 6 for further details).

Physically, caterpillars are elongate, fluid-filled bags that come in a variety of shapes. Some are sluglike, others are weiner-shaped. Some are smooth while others are covered with spines, fleshy *tubercles*, or irritating hairs. Each caterpillar possesses a distinct head equipped with minute *antennae*, chewing mandibles, and six pairs of simple eyes or *ocelli*. A triangular piece, the *frons*, separates six

10. Some larvae, such as those of the Monarch butterfly (*Danaus plexippus*) are elongate and smooth. The fleshy "tentacles" are not poisonous. (After Scudder 1889.)

ocelli on each side of the head. Closer examination reveals that the head has evolved from an ancestral form with distinct segments. Each segment formerly bore a pair of jointed appendages. Over millions of years these segments were coalesced to form what appears to be a single functional unit (the head), while the jointed appendages of the former segments were modified to form head appendages such as the antennae, as well as complex mouthparts for handling food.

The first primitive head section bears two hardened wedge-shaped mandibles used for crushing the food plant. A *labrum*—an appendagelike structure not derived from a segmented appendage—lies above the mandibles and acts like an upper lip. Minute antennae composed of three segments lie on either side of the mandibles. Behind the mandibles (and derived from the appendages of the next primitive segment) are the maxillae with their sensory *maxillary palps*. The last head segment bears the *labia*, a kind of lower lip, which together with the maxillae helps to hold food between the mandibles for chewing.

The *spinneret*—a silk-spinning spigot and its associated silk glands—is composed partly of structures derived from the maxillae, a fleshy *hypopharynx* (not derived from a walking leg), and the labia.

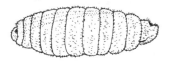

11. The larvae of many lycaenid larvae such as the Great Copper butterfly (*Lycaena xanthoides*) are slug-shaped and flattened. They are extremely difficult to find on their food plant. (After Scudder 1889.)

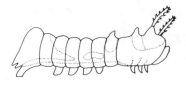

12. The larvae of the tree-feeding Red-spotted Purple butterfly (*Limenitis arthemis astyanax*) appear to be small knobby sticks. (After Scudder 1889.)

Thus, the larval spinneret lies between the maxillae and the labium. Silk is really a complex protein whose composition may vary within and between different families of butterflies. The *labial silk glands* manufacture the silk and pass it to a common duct where it is squeezed and flattened into a fine ribbon by muscular contractions. Silk consists of about 75 percent fibroin, an elastic and strong protein, and it is used for a number of purposes (Wigglesworth 1972).

The remainder of the larval body consists of thirteen segments—three thoracic segments bearing one pair each of jointed walking legs with a single terminal claw, and ten abdominal segments, five of which are equipped with fleshy, nonsegmented larval legs, or *prolegs*. The prolegs have a series of ventral hooks, or *crochets*, for grasping.

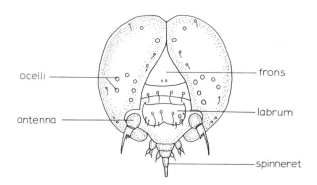

13. This front view of the larval head of the Eastern Tailed Blue butterfly (*Everes comyntas*) shows the major structural features. (After Lawrence and Downey, in Howe 1975.)

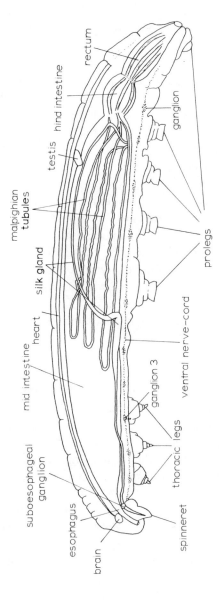

14. Diagrammatic view of the major internal organs of a male Monarch caterpillar. The digestive system takes up the majority of the interior. The open heart-circulatory system lies above the digestive system while the solid ventral nerve cord lies below. Note that the silk glands extend back from the mouth well through the larval interior. (After Scudder, in Howe 1975.)

Except for the head capsule, the external skeleton (*exoskeleton*) of the caterpillar is not as heavily *sclerotized* or hardened as that of an adult butterfly. Instead, it is the pressure of the living tissue and body fluids within that keep the wormlike body of the larva inflated. If the cuticle is punctured, the *hydrostatic exoskeleton* oozes fluid and collapses. A large puncture in the exoskeleton will quickly deflate and kill the caterpillar. However, ventilation does not deflate the caterpillar because gases are exchanged through a row of thoracic and abdominal spiracles, or openings, each with its own valve. These spiracles lead to long air tubes (tracheae) that ramify throughout the entire body, and penetrate all organs, even small muscle fibers.

The caterpillar's duty is to eat—it is a food-processing machine of formidable capacity. So much food is consumed so rapidly that the caterpillar gains weight at nearly an exponential rate. As a consequence of this singular purpose in its life, the caterpillar's internal "control system" is about as straightforward as any electrical engineer could wish. From a minute brain leads a paired, ventral nerve cord studded with pairs of knotlike masses of nerve cells called ganglia in each body segment. The nerve cord and its ganglia run the length of the body.

Equally simple is the elongate digestive bag that contains distinct compartments along its length. Spaghettilike *malpighian tubules* undulate loosely around the posterior of the digestive system, suspended within the body cavity and bathed by body fluids. Their primary function, like that of our kidneys, is to eliminate nitrogenous wastes—largely uric acid with trace amounts of urea. Malpighian tubules also help balance the ionic composition of the body fluids by regulating the concentrations of salts, sugars, and amino acids.

Another larval feature of note is a dorsal, tubular "heart" that receives blood, or haemolymph, from the body cavities through a posterior opening, as well as through segmentally arranged paired holes termed *ostia*. The heart then pumps the haemolymph anteriorly toward the head and anterior body where it is shunted back into other body regions. The caterpillar's circulatory system is called an "open system" because it has no truly closed, continuous vessel system like that of vertebrates.

The thorax bears three pairs of jointed legs. These legs are used

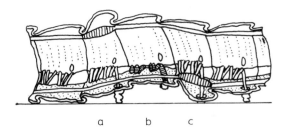

a   b   c

15. The sequence of muscular contraction and relaxation are shown in this longitudinal section of a hypothetical caterpillar. Segment *a* is being shortened by the contraction of the dorsal longitudinal bands, as the dorsal-ventral muscles and those operating the proleg of *b* and the ventral longitudinal muscles of *c* are contracted. (After Barth 1937.)

partly to help manipulate food, and together with the five pairs of fleshy prolegs and their grasping crochets, they are also used to crawl—in caterpillar fashion, of course. The caterpillar crawls by shortening and lengthening the abdominal segments in a wavelike fashion, beginning with the last or anal abdominal segment. As the anal segment is compressed and then pushed forward, the next-to-last segment is also compressed and then extended, passing the compression wave to anterior segments.

The muscles that provide for locomotion are not the many small bands extending just beneath the skin folds—they assist in maintaining the hydrostatic pressure: if a caterpillar is pierced, these muscular rings quickly contract around the puncture site, sealing it off and preventing the larva from "leaking to death." Instead, the muscles responsible for a caterpillar's crawl are found in each segment and consist of long muscles—both *dorsal longitudinal* and *ventral longitudinal* muscles—as well as a series of shorter *dorsal-ventral muscles* (Wigglesworth 1972).

A proleg is lifted when the dorsal longitudinal muscles in the segment anterior to it and the ventral longitudinal muscles in the segment behind it contract. This action buckles or puckers the anterior segment at the top and the posterior segment at the bottom. Simultaneously, the dorsal-ventral muscles contract, thereby releasing the suctionlike grasp of the proleg on the stem or leaf rib. When these same muscles perform the opposite function, contracting and relaxing, the compression wave is passed onward and the proleg is lowered. The flat bottom of the proleg strikes the surface and the central area is drawn up, creating a small suction, while the crochets dig into the substrate.

So, the caterpillar moves through a series of antagonistic muscle contractions, creating a pressure wave in the last segment that passes anteriorly. If the larva were flaccid, not enough pressure could be generated. Because it is turgid, the fluid pressure passed forward to each successive anterior segment is relatively large, allowing the largest caterpillars to exert considerable strength against a substrate. If it is necessary to move a large caterpillar quickly, one good way is to grasp it firmly and then pull the whole animal away in one smooth but rapid pull. Pulling slowly allows more crochets to be engaged with the result that the exoskeleton is torn in many places.

THE METAMORPHOSIS

Despite an incredible amount of external variation in form and color, caterpillars share one important physiological process—*molting*. Molting is the process whereby the old, relatively nonexpandable exoskeleton is partly digested by enzymes, then recycled below the old exoskeleton to form a new, larger exoskeleton. At the end of the molting process a remnant of the old exoskeleton is shed, and the new, larger exoskeleton expands to accommodate further growth. The larval form between molts is called an *instar*, commencing with the first instar, between hatching and the first molt, followed by the second instar, between the first and second molt, etc.

The molting process is a very complex series of biochemical reactions under neural and hormonal (neurosecretory) control. In most insects there are stretch receptors or neural sensors located in strategic positions that "fire" when the exoskeleton is stretched to capacity. The response of these stretch receptors is relayed to the brain along with physiological responses due to other external and internal stimuli, all of which inform the caterpillar of its state of "molt readiness." A brain hormone is then released that mediates the production of other hormones. The brain hormone flows down the nerves from the neurosecretory *pars intercerebralis*, and is stored in an area of the neurohaemal area of the brain termed the *corpora cardiaca*.

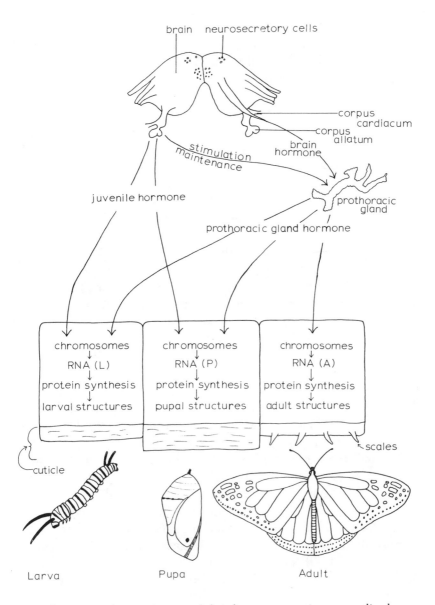

16. The major endocrine tissues and their hormone secretions are outlined on this diagram, illustrating how the varying concentrations of the different hormones affect metamorphosis in the Lepidoptera. (After Gilbert 1964, in Atkins 1978.)

The brain hormone then enters the haemolymph or directly stimulates the *prothoracic glands*, which produce a mixture of related molting hormones generically called *ecdysone*. The ecdysones stimulate the division of epithelial cells, and it is their secretions that form the new exoskeleton. If the next stage is to be another larval instar, then the *corpora allata*—a minute pair of neurosecretory lobes in the brain—secrete large quantities of *juvenile hormone* into the haemolymph, which stimulates the epithelial cells into maintaining the larval "status quo." However, if the next stage is to be the chrysalis, the molting hormone is joined by only a minimal amount of juvenile hormone. Finally, if the pupa-adult transition is the next stage, virtually no juvenile hormone is produced, and only the molting hormone and another *eclosion* hormone are produced by the neurosecretory cells in the brain.

Basically, then, it is the quantity of juvenile hormone that determines whether the next stage will be a larva, pupa, or adult. This is because juvenile hormone in the haemolymph somehow suppresses the expression of those genes responsible for adult characteristics. If the brain of an immature insect is removed after the larva has fed for a few days, molting activity ceases because no brain hormone is present to maintain the integrity of the prothoracic glands, and hence these glands cannot manufacture the molting hormone ecdysone. If the flow of haemolymph is constricted by tying the body between the head and thorax, and between the thorax and abdomen, and then brain tissue is transplanted into each section (thoracic and abdominal) the thorax will molt, but the abdomen will not since ecdysone cannot reach the abdominal tissues (Wigglesworth 1972).

An extraordinarily large butterfly can be created by performing the following microsurgical transplant: take a late instar larva that ordinarily would molt into a chrysalis, and transplant into it the corpora allata from an early instar. The caterpillar with the corpora allata removed will develop into a tiny adult butterfly because it cannot produce any juvenile hormone—the hormone that blocks the expression of adult characteristics.

The caterpillar with the transplanted corpora allata, however, will produce too much juvenile hormone and will be unable to molt to the pupal stage. Instead, it will molt to a "super" caterpillar. This giant larva may ultimately molt to a gigantic chrysalis that produces a huge adult. The reason for these bizarre developments is that even

first instar caterpillars have tiny *imaginal discs,* or groups of immature cells, with the potential to develop into adult tissues. As soon
as the production of juvenile hormone falls below a minimal level,
metamorphosis can proceed regardless of the size of the larva.

During the last instar, the caterpillar gains a phenomenal amount
of weight and increases considerably in size. Then, in anticipation of
the spectacular last molt from larva to pupa, the caterpillar often
defecates a syrupy, sticky "last frass" and may even change color as
the pupal skin begins to form under the larval exoskeleton. The
caterpillar may roam considerable distances before choosing an appropriate, usually protected site for pupation.

THE CHRYSALIS

Just prior to pupation, most caterpillars spin a silk button or pad on
the substrate surface. Pierid, papilionid, and lycaenid larvae also
typically spin a U-shaped silk girdle around the first or second abdominal segments. This guidewire acts like a lineman's safety belt.
Caterpillars that construct a girdle usually position themselves fac-

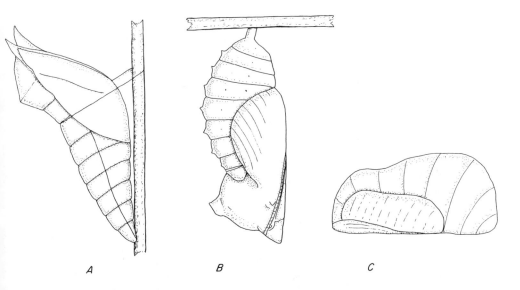

A *B* *C*

17. The different shapes of pupae: *A,* Papilionidae; *B,* Nymphalidae; *C,*
Lycaenidae.

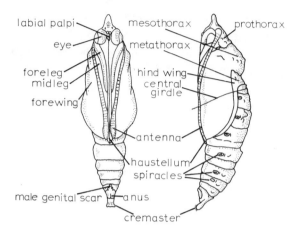

18. The major external features of a butterfly chrysalis. Note the distinct encasements (although fused to the chrysalis proper) for the legs, the wings, and the sense organs and mouthparts. (After Emmel 1975.)

ing upward or horizontally, rarely downward. Most other butterfly larvae hang downward and do not spin a supporting girdle. As always, however, there are exceptions. Primitive butterflies such as *Baronia brevicornis* and *Parnassius* spp. respectively pupate underground in loose soil, or construct crude cocoons in the leaf litter on the surface (Tyler 1975).

All other pupae sink a series of special hooks on the last abdominal segment—the *cremaster*—into the silk pad just as the last of the larval exoskeleton slides over the abdomen. The actual "stripping" of the larval skin at the end of a molt is termed *ecdysis*, an appropriate description. As the larval skin splits over the head of the new chrysalis, the wing pads, legs, antennae, and mouthparts are at first almost free of the body. But within minutes these casings join the pupa proper and hours later the chrysalis appears as a single cohesive structure, even though the various appendages and their individual cases can be easily distinguished.

The chrysalis is a stage of metamorphosis that is exposed to much adversity: intense cold, severe rainstorms, or prolonged droughts undoubtedly destroy millions. Furthermore, there is practically no defense from predators and parasites except for camouflage, and the ability to maintain a lifeless position. Even then, ants can cause considerable mortality in tropical butterflies such as *Heliconius ethilla* (Ehrlich and Gilbert 1973).

Sometimes the emerging pupa falls because the hooks of the cremaster have not been securely fastened to the silk pad. In those species without a silken girdle, the chrysalis falls to its death. Even if the pupa survives, the butterfly may be crippled because it cannot cast off the pupal shell during eclosion—there is nothing to pull against and remove the pupal skin from the body.

If the pupa has a girdle and the cremaster hooks miss while the pupal case is still soft, the silken girdle cuts deeply into the surface, often into the interior of the pupa, causing developmental deformities. Yet, despite all the potential hazards—overexposure to sunlight, drought, wind, lack of sufficient food reserves, all of which can prevent successful eclosion—pupation seems to succeed most of the time.

The chrysalis stage is often erroneously called a resting stage, but nothing could be further from the truth. Larval features are dismantled chemically and embryonic cells divide. After only a few hours, the adult features begin to form: minute wing stubs, mouthparts, thoracic muscles, and legs. This, of course, occurs in pupae that do not diapause. The imaginal discs within diapausing pupae are "turned off" until environmental signals such as higher humidity (in subtropical areas subject to seasonally predictable droughts), or warmer temperatures and longer day lengths (in temperate regions) trigger adult development.

The pupa is the metamorphic bridge between larva and adult. Some epidermal cells are dying while others are actively dividing. Many larval features are dismantled as imaginal cells replace those of the larva with those of the adult. In butterflies, the adult cuticle is laid down by the same epidermal cells that formed the larval exoskeleton. The entire leg of the larva becomes germinative tissue and differentiates to form the complicated structures of the adult leg. Generally then, there is a progressive substitution of adult cells for larval cells. As larval structures disappear, they are gradually replaced by adult structures arising from the imaginal discs (Wigglesworth 1972).

The mystery of metamorphosis lies in its control. During the pupal stage, genes controlling the expression of larval characteristics must be turned off while those for adult characteristics

must be turned on. Yet, the same genetic constitution, or genome—
for both larval and adult characteristics—has been present in each
individual since fertilization. In a way then, a butterfly is *polymorphic* as well as metamorphic over the course of its lifetime. That
is, it has more than one discrete superficial form.

The word *polymorphism* enjoys a wide and complex usage today.
It may refer to discrete, discontinuous differences within an individual, or differences between individuals—regardless of whether
the polymorphism takes place at the gene, enzyme, or organ level. A
phenotypic polymorphism is usually genetically induced, but commonly moderated or enhanced by environmental factors such as
temperature, light intensity, or even behavioral conditions like
crowding (Downey 1965a). Chapter 7 treats larval, pupal, and adult
polymorphism and related phenomena in more detail.

Superficially then, the stages of metamorphosis represent what
appears to be a discrete polymorphism. During the course of evolution, a line of insects succeeded in genetically suppressing all adult
characteristics during the larval stage, particularly the developing
wing pads of the adults. These so-called holometabolous insects
evolved the transitional stage (pupa) to bridge the morphological gap
that developed as the ecological niches of larvae and adults diverged.
Other lineages of insects (e.g., grasshoppers) undergo a gradual or
hemimetabolous metamorphosis in which adult characteristics
(such as mature sex organs and functional wings) are acquired slowly
through a succession of instars.

In the holometabolous butterflies, the chrysalis is the metamorphic stage between larva and adult. However, organs and major
structures like leg appendages are already undergoing a primary re-
organization even in early larval instars. These differences are internal and therefore cannot be documented by external observation.
More radical internal changes then take place at a more rapid pace in
nondiapausing pupae. For example, the order of the veins and the
shape of the pupal wings are determined when the pupa is between
twenty-four and forty-eight hours old. And wing colors are formed
from genetically determined focal points on each wing, even before
the minute scales have formed (Nijhout 1981).

In fact, there is a succession of critical periods in wing development that can be affected by environmental conditions, especially
temperature and day length. When pupae of the Common Tor-

toiseshell (*Aglias urticae*) are exposed to heat during these successive critical periods, patterns and colors on different parts of the wings are affected. If pupae are exposed to temperatures of about 49°C (120°F) for short periods during the first forty-eight hours after pupation, the wing pattern is modified. If exposed to high temperatures during the following twenty-two hours, scales may be lost, and if exposed between the subsequent ten hours, the form of the scale is affected. These differences occur because wing scale development and pigment production do not occur precisely in tandem. The development of wing pattern and pigmentation is discussed more fully in chapter 3.

When development is complete, a seam along the front of the pupal exoskeleton is digested by enzymes, then ruptured by the imago pushing outward with its head and expanding thorax. The wet, crumpled butterfly grasps the pupal skin or any convenient perch, and once satisfied with its position, proceeds to pump the wings full of haemolymph, thereby expanding them to their functional size.

All in all, pupae may seem rather uninteresting. However, even pupae can exhibit fascinating structures and behaviors. For example, the pupae of many lycaenid species produce tiny creaking or chirping noises. Even the well-known Monarch butterfly (*Danaus plexippus*), can produce audible clicking sounds in the pupal stage (Downey 1966), a fact that is little appreciated.

There are several ways in which pupae can produce sound: (1) the body can be slammed against its supporting structure as do the gregarious pupae of the Mourning Cloak (*Nymphalis antiopa*) (see also chap. 6), (2) one or more pairs of abdominal segments can be rubbed together where their margins meet, and (3) the ventral portion of the abdomen can be rubbed against the case enshrouding the proboscis. Over 150 species of lycaenid and riodinid pupae produce chirping, creaking, clicking, and even humming noises by the second method. The microscopic rasp and file, or *stridulatory organ*, may encircle the abdomen completely, but is usually confined to the dorsal side of the chrysalis. The file and stridulating plate, typically located between the fourth and fifth abdominal segments, or the fifth and sixth, are ground together. The speed and force with which this rubbing action takes place affects the loudness and pitch of the sound emitted (Downey 1966).

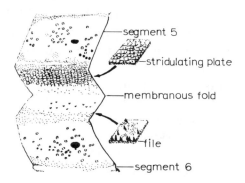

19. The stridulatory structures of a lycaenid pupa are shown in these partially diagrammatic drawings of a chrysalis surface, showing the dorsal view of membrane 5 in the region of the spiracle. Inserts show the enlargement of the file in the chrysalis of the Mormon Metalmark (*Apodemia mormo virgulti*) that produces the sound as the file is drawn across the stridulatory plate. (From Downey 1966.)

The stridulatory plate consists of a heavily sclerotized section whose surface is regularly pitted or roughened. Although the stridulatory plates are somewhat similar in size and shape in closely related species, the grating area may be composed of various surfaces: pimplelike or knoblike tubercles; a series of granular prominences uniformly arranged but irregular in size like grainy leather; or ridges with crests and two or more sloping surfaces. The opposing file typically consists of sharply pointed, conspicuous teeth. Muscles attached to the membranous areas in the fold between the segments move the file and stridulatory plate against each other to produce the sound (Downey 1966).

Some species such as the Bronze Copper (*Lycaena hyllus*) produce two types of sounds—a distinct chirp and a slight humming noise. Strangely, the sounds are produced when pupae are agitated externally, say by tapping them, but seldom spontaneously. Why are the pupal sounds produced? No one really knows for certain. A defensive function is most commonly ascribed, and this makes some sense since some singing pupae stridulate when disturbed. Perhaps pupae squeak when they are about to eclose, a formal way of announcing a "coming out" for potential mates that may be in the area.

Yet others point out that butterflies rely largely on visual and chemical cues to attract mates (for a complete discussion see chap. 7), and that the pupal sounds are not only scarcely audible, but identical in both sexes. Furthermore, sound production can occur at any time after the pupa is formed, and no acoustical receptors have been found in any pupae. Perhaps though, ultrasonic frequencies are also involved that are beyond the capacity of the human ear (Downey 1966).

Another, perhaps more sound explanation for pupal stridulation concerns the unique symbiotic relationship involving many species of lycaenid butterflies and their attendant ants, which milk the caterpillars for their special sugary secretions. The secretions ooze from tiny glands located on the seventh abdominal segments. Rarely, they may function in the pupal as well as in the larval stage. Unfortunately, many species with stridulatory organs have never been found associated with ants. Of course, the lack of an association with ants may represent a modern, derived condition, and their ancestors may have been formerly associated with ants (see chap. 6).

For example, the pupae of some Australian species emit sounds only in the presence of ants and are otherwise silent. Furthermore, researchers report that ants also stridulate, and stridulation is an important means of communication in these highly social animals. Thus, lycaenid pupae may advertise their presence to ants by chirping or humming noises. Once attracted to the pupae, the ants may be drawn to the sugary spigots, a food resource they will protect from potential predators and parasites (Downey 1966).

But there are problems with this interpretation also, since the Harvester butterfly (*Feniseca tarquinius*)—a species that is predaceous on aphids and mealybugs and is *not* associated with ants in the larval stage—has well-developed stridulatory organs, whereas the Silvery Blue (*Glaucopsyche lygdamus*), attended by at least three species of ants, has vestigial stridulatory organs and produces no sound! Downey concludes that the lycaenid stridulatory organs may in part be evolutionary legacies present in the ancestors of lycaenid butterflies that formerly assisted the pupae during eclosion. The same muscles that produce sound also permit abdominal flexion and the pushing action necessary to rupture through the confining pupal case. Perhaps, then, the muscles gained an additional evolutionary advantage in being used for sound production. To date, however, there is no consensus on the function of stridulation, and clearly, there remains a great deal of research to be done with these "noisy" pupae!

CHAPTER 3

The Imago

Under summer conditions the transformation from pupa to *imago* (adult) can take anywhere from four days to two weeks. However, under extreme environmental conditions pupae may diapause as long as one or two years before metamorphosis is complete. As the imaginal discs differentiate to form the intricate structures of the adult butterfly, the color of the pupa begins to darken. Normally during this tissue formation, or histogenesis, very little water is lost by respiration. This is largely because the *chitin* of the exoskeleton is thick, not easily punctured, and resistant to the passage of water both into and out of the pupa.

As the metamorphic transformation nears completion, the pattern and color of the butterfly's wings become apparent through the pupal cuticle. When the pupal shell is quite brittle, so that the slightest touch produces an audible crumpling sound, the *pharate* (hidden) adult is ready to emerge, or eclose. Inhaled air and the expansion of the muscles in the thorax split the exoskeleton near the apex of the pupa. As the head with its antennae pull away from their cuticular compartments, the legs quickly follow and grasp for the first perch available—often the pupal skin itself. Within seconds, the abdomen is freed and the fine internal linings of the tracheae are everted just as would happen if you pulled your hand out of a sticky rubber glove.

The damp insect with its swollen abdomen then secures a permanent perch, usually upside down, and typically after much seemingly aimless crawling and maneuvering. Sometimes, in the scramble to find a perch, the butterfly falls, and the fragile wings ooze haemolymph as the wing veins rupture. These butterflies, like those that fail to shed the entire pupal skin, are destined to be crippled.

Cripples cannot fly, and they quickly die from predation, exposure to the elements, or lack of nourishment. If all goes well, the new butterfly, hanging from its perch, begins a series of rhythmic contractions of the abdomen, up and down, pumping the haemolymph through the wing veins. With each contraction the wings slowly unfurl, until finally the butterfly hangs fully extended, limp yet spectacular in its garden-fresh beauty.

During or immediately after expansion of the wings, the butterfly performs several vital functions. The first is the elimination of *meconium*, a liquid waste that is sometimes brightly colored. In communal butterflies like the Mourning Cloak (*Nymphalis antiopa*), the bright red meconium is forcibly ejected as a fine spray. Since eclosion is often synchronous in this species, any disturbance of the pupation site may cause dozens of freshly emerged butterflies to stain the intruder a bright red. So although the meconium is certainly a waste substance, in some species its ejection assumes a defensive purpose as well, even though it may not be toxic. A second important activity is to clean the antennae by dragging them through a crook of the forelegs, and a third, most important operation is fitting the two zipperlike halves of the proboscis together so that it forms a functional, strawlike appendage.

Many of the structures of the adult butterfly are important in classification. Because butterflies cannot always be bred easily to establish their genetic compatibility, lepidopterists, like most entomologists, must rely upon traditional ways of distinguishing one species from another. This is usually accomplished by comparing physical characters: wing patterns, colors, and the morphology of the sexual organs.

However, this method has caused problems in the past since characters (e.g., eyespots) and their "states" (e.g., blue eyespots versus green eyespots) may vary continuously rather than discretely as we would like. The result has been a proliferation of specific and subspecific names, as well as forms, aberrations, and varieties, many of which are not scientifically valid. Because the adult characters and their states are far more complex and better known than those of the egg, larva, or pupa, the remainder of this chapter is devoted to their brief description.

In the reductionist view, the imago is nothing more than a flying sex organ and dispersal agent, albeit a gaudy one. Yet, despite the

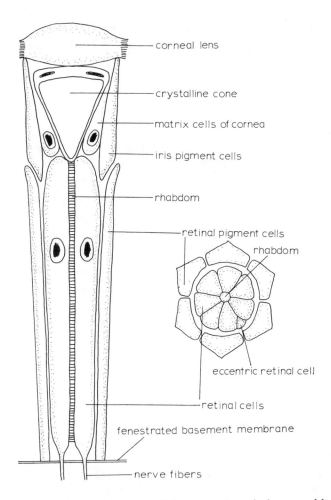

corneal lens

crystalline cone

matrix cells of cornea

iris pigment cells

rhabdom

retinal pigment cells

rhabdom

eccentric retinal cell

retinal cells

fenestrated basement membrane

nerve fibers

20. Diagram of the apposition eye. Light passes through the corneal lens and the crystalline cone, down through the rhabdom formed by the retinal or sense cells, and ultimately stimulates the optic nerve. Cross stimulation of adjacent ommatidia is probably prevented by the retinal pigment cells that surround the rhabdom. (After Snodgrass 1935.)

flashy wings, the adult body is similar in some respects to that of the larva. For example, both stages possess three distinct sections, or *tagmata:* head, thorax, and abdomen. However, each of these adult tagmata is more complex than those in the larval state.

The head is really the feeding station and computer terminal for much of what a butterfly senses about its environment, from motion to smell. The huge compound eyes are composed of two to twenty thousand transparent facets called *ommatidia*. A bubble-shaped, transparent, and hexagonal *corneal lens* allows light to pass via a crystalline, conelike structure down through a tubular optical rod or *rhabdom* secreted by eight long retinuli cells arranged in an octagon. Each ommatidium apparently receives and transmits information about the light intensity from one point of the visual field. Pigment cells surrounding the crystalline cone may prevent adjacent ommatidia from being stimulated by more than one source of light. The result is a mosaic picture, much like a computer printout composed of dots.

The resolution of the eye depends partly on the number of ommatidia, their orientation as a hemisphere of adjacent lenses, and the development of the optic lobe—one of the largest areas of the brain. Visual acuity may also vary with the sex of the butterfly. The eyes of males are generally larger and may be better than those of females because male eyes have more ommatidia. This would not be surprising, since most males use visual cues to locate mates (Silberglied 1984).

While the acuity of butterfly vision can hardly approach that of the vertebrate eye, it is sharp enough to locate mates, food resources, and appropriate larval food plants, as well as avoid avian predators—at least most of the time. In addition, butterflies can see wavelengths that we cannot see, such as ultraviolet and infrared. This gives them the ability to see "colors" in these wavelengths where our eyes are not sensitive (Swihart 1972; Bernard 1979).

In fact, butterflies and many species of bees are attracted to certain flowers on the basis of their ultraviolet color patterns. Furthermore, some butterfly species are sexually *dimorphic* when viewed under these wavelengths: males and females that look alike under visible light differ considerably under ultraviolet light. For example, the wings of male Orange Sulphur butterflies (*Colias eurytheme*) are brightly metallic and reflective under ultraviolet light whereas those of the female are orange in the visible and dark under ultraviolet wavelengths. Thus, wing color may differ under different wavelengths, making color important in mate identification (Silberglied and Taylor 1973) (see chap. 7 for a complete discussion).

Butterfly eyes have other peculiar physical attributes. One of

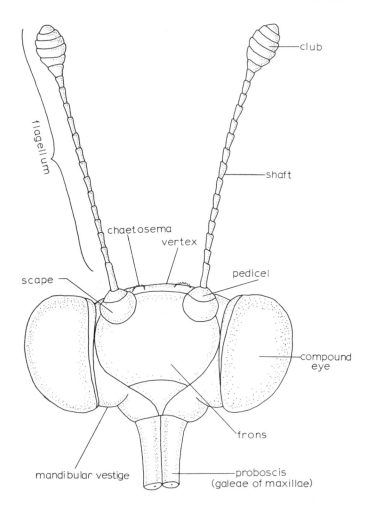

21. The major external features of the butterfly head. For clarity, the labial palpi and head scales are not shown. (After Howe 1975.)

these is the presence of reflective spots that seem to change their position when the butterfly eye is viewed from different angles (Sibatani 1973). The first detailed description of these spots was published by two Japanese scientists (Yagi and Koyama 1963). Since the presence of the reflective spots is apparently due to the morphology of the ommatidium, butterflies can be classified to a certain taxonomic level on the basis of their reflective eyespot patterns.

The basic pattern is a large, central reflective spot surrounded by six minor reflective spots. In some cases these minor spots are surrounded by another series of six to twelve even smaller spots. The reflective spots are thought to be caused by the partial absence of pigmentation in the cells of the distal half of the ommatidia. A lack of pigmentation would allow incident light from one ommatidium to pass through obliquely and strike adjacent ommatidia, thereby causing minute points of reflection. Dark eyes without the reflective spots appear dark probably because the ommatidia are optically separated from each other. But why the reflective spots should be distinctive in living insects and then gradually disappear after death remains an unanswered question.

Behind the compound eyes and near the top of the head lie two *chaetosemata*—small bumps covered with many stiff hairs that possibly serve as sensory organs, although their exact function is unknown. Just below the chaetosemata at the upper corners of the triangular frons project two boldly knobbed antennae, the primary sensory organs of the adult. The antennae contain many different types of *sensilla*—sensory cells that respond to vibrations, sound, and chemical stimuli such as airborne odors.

Each antennae is composed of a donut-shaped basal *scape*, followed by a short, wristlike *pedicel* that leads to a many-segmented *flagellum*. The flagellar segments are approximately equal in width, but begin to flare in diameter near the tip so that the antennae terminates in a distinctive knob or club. Like most of the body, the head and the antennae are partly covered with hairs and their flattened counterparts, the *scales*. The *nudum* is the general name given to the antennal area devoid of both scales and hairs, although the nudum varies in size and shape from species to species (Clench 1975).

The antennae also house sensilla that respond to chemical sex signals—the *pheromones*. These pheromones play a significant role in butterfly courtship and mate recognition (Grula and Taylor 1979), perhaps more so than the initial visual cues. For example, Silberglied and Taylor (1978) have shown that mate selection by female Orange Sulphur butterflies (*Colias eurytheme*) requires both visual and pheromonal cues while that of the Common Sulphur (*C. philodice*) requires only pheromonal cues.

Grula has examined the antennal sensilla of these two closely

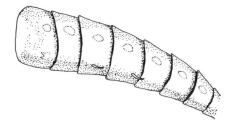

22. One oval or roundish depression is present in the middle of the terminal eight subsegments of the *Colias* antenna. (From Grula and Taylor 1980a.)

related sulphur butterflies in an attempt to locate chemically sensitive sensilla. He describes three basic types of sensilla, two of which are likely to be important in detecting olfactory—including pheromonal—stimuli. Unexpectedly, however, there was little if any sexual dimorphism, or discrete sexual differences, in the minute structures and sensilla of the antennae, and furthermore, the gross and fine antennal features of both species and both sexes were nearly identical (Grula and Taylor 1980a).

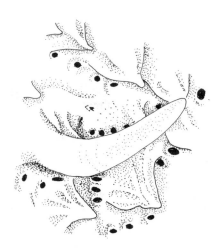

23. Numerous pores are visible in this close-up of a curved, thin-walled peg. Such pegs on the *Colias* antenna are surrounded by even smaller microtrichia. (From Grula and Taylor 1980.)

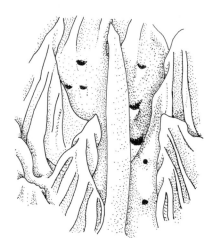

24. Microtrichia also surround the short, thin-walled pegs found in the depressions of the *Colias* antenna. (From Grula and Taylor 1980a.)

Perhaps there are differences in the nerves that activate the sensilla, or differences in the areas of the brain controlling the activity of the antennae and their responses to external stimuli? Such sexual differences in brain morphology are certainly well documented in mammals, including humans, so it would not be a complete surprise to discover similar neural differences in the brains of butterflies, especially since many species exhibit profound sexual dimorphism (see chap. 7).

In sulphur butterflies, each flagellum is composed of twenty-seven segments, and most of the olfactory sensilla are located on the inner or medial surface of the nudum. Every segment after the fifth or sixth segment from the pedicel contains an oval or round depression bearing curved, thin-walled pegs with perforations. Minute hairs, or *microtrichia*, cover the remainder of the antennae. The thin-walled pegs with their perforated surfaces supply numerous microscopic pores, possibly for the entry of airborne scent molecules, including pheromones.

Colias antennae also bear approximately four hundred long, thick-walled hairs with grooved surfaces that lack perforations. These thick-walled hairs may be important in chemical reception, but more likely perceive touch and mechanical distortion, such as

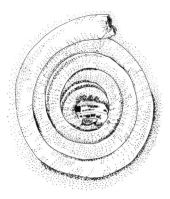

25. A scanning electron micrograph of a *Danaus* proboscis shows a lateral view of its tubular, but tightly coiled structure. (From Kingsolver and Daniel 1979.)

occurs when the antennae bends to touch some object. Similar antennal sensory hairs have been found on the Florida Queen butterfly (*Danaus gilippus berenice*) (Myers 1968; Myers 1969; Myers and Brower 1969).

Projecting downward and anteriorly from the triangular face (frons) is a medial proboscis, the hollow, strawlike sucking tube bounded on either side by *labial palps*. The proboscis is derived from the galeae—two lancelike projections of the maxillae. As some have suggested, the proboscis may have first evolved as a lapping structure, or a structure for the extraction or removal of water (Neck 1978), since some butterflies use the proboscis to remove excess dew from the wings. In my view, however, water removal is more likely a function that has evolved subsequent to the development of a tubular proboscis for the procurement of liquid food.

Whatever the reason(s) for its development, the proboscis of contemporary butterflies is really composed of two maxillary projections that have been greatly elongated. The two halves are zipped together by rows of teeth both dorsally and ventrally, forming the hollow food channel. This channel terminates in a *cibarium* or anterior suction cavity. If not in use, the prehensile proboscis is coiled tightly like a watch spring.

When nectar, water, or other fluids are pumped by the action of

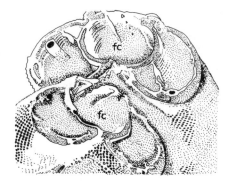

26. The food canal (fc) of a *Colias* proboscis is shown in this cross sectional view. Note the toothed margins making up the left and right halves of the canal. (From Kingsolver and Daniel 1979.)

the two sets of muscles that dilate the cibarium, the bulblike cavity expands, producing a vacuum. The energy needed to draw nectar through the proboscis is directly proportional to the pressure drop produced by the cibarial muscles. However, the rate of nectar extraction depends upon the thickness or viscosity of the nectar, the cibarial pressure drop, and the length and diameter of the proboscis (Kingsolver and Daniel 1979).

The adult butterfly has chemical receptors in the *tarsi* of the legs, and when these small terminal subsegments are stimulated, their response encourages the butterfly to probe for nectar. Additional sensors are located in the tip of the proboscis. If a butterfly is placed near a drop of honey on a wet towel, the proboscis is used as a fine probe, gently tapping near the spot of honey in order to find the highest concentration of sugar that can be efficiently withdrawn without overtaxing the physical limits of the proboscis. However, if the honey and water solution is mixed evenly before spreading it on the towel, the butterfly will remain in one place without extensive probing.

Kingsolver and Daniel (1979) have made some interesting observations concerning the operation of the proboscis. Although some butterflies feed on urine and liquid excrement, tree sap, and even decaying fruits and animals, most rely on the sugars in nectar as an energy source (see chap. 4). Nectar is largely composed of single and double sugars—mono- and disaccharides respectively—dissolved in water, and sometimes spiced with minute amounts of amino acids (Watt, Hoch, and Mills 1974; Baker and Baker 1975). This means that nectar is largely a watery solution, perhaps fit for the gods and butterflies, but not for mankind as a staple.

Kingsolver and Daniel (1979) assume that nectar flow through the proboscis is not turbulent, and that the streams of nectar flow in discrete layers past each other within the proboscis. This "layering of molecules" in a fluid keeps turbulent eddies and backcurrents to a minimum. Kingsolver and Daniel also assume—since there are no large molecules in nectar—that the fluid is Newtonian. This means that the thickness or viscosity of nectar is independent of movement of the individual fluid layers, regardless of the temperature of the fluid.

Capillary action is apparently of minimal importance in drawing liquid food up the proboscis because it would function well only when the food channel is filled with nectar. Since the flowers on which butterflies feed typically have one hundred times the volume of the average (e.g., *Colias eurytheme*) food channel in the proboscis, capillary action is simply not efficient at transporting nectar upward.

The theoretical simulations done by Kingsolver and Daniel (1979) show that nectar extraction is limited by the mechanical constraints imposed by the proboscis and its cibarial pump. The optimum for the greatest net power gain or energy extraction rate in the feeding system of the butterfly occurs with a fluid having a 20 to 25 percent sugar concentration. This concentration range overlaps that found in butterfly-pollinated flowers, which typically have a 15 to 25 percent sugar concentration (Watt, Hoch, and Mills 1974). It seems, then, that butterflies apparently have been honed by physical restraints to feed on flowers that produce nectars with a sugar concentration between 15 and 25 percent, thereby maximizing the net power gain of the system.

Optimal power gain is possible when butterfly temperatures are low and flower temperatures are high. However, most butterflies need high body temperatures (above 20°C or 70°F) to fly efficiently (see chap. 4). Furthermore, high flower temperatures might increase the evaporation of water, thereby increasing the concentration of the nectar and making it too syrupy to transport efficiently up the proboscis.

So when a *Colias* butterfly extracts nectar at a rate of 0.5 micro-

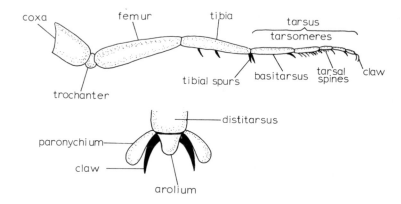

27. The legs of butterflies are composed of five distinct segments. In the case of the Black Swallowtail (*Papilio polyxenes asterius*), the tarsus is composed of five subsegments with tarsal spines that terminate in a distitarsus with claws, arolium, and pads (paronychia). (After Clench 1975.)

liter per second (one microliter is one millionth of a liter) from a flower containing 0.5 to 5.0 microliters of nectar, it is doing rather well. Butterflies may spend anywhere from several seconds to almost a minute probing a given flower, but the visitation period must vary considerably in accordance with other physical and behavioral variables about which we know little.

Like the head and abdomen, most of the thorax is covered with fine hairs and scales. Unlike the head, however, the thorax is composed of three distinct segments, each bearing a pair of jointed legs. The legs likely were the original and only set of locomotor structures in the ancestral wingless insects. An adult's legs develop from the larval legs in such a way that each larval leg acts like one huge imaginal disc. Thus, all larval leg sections are required to produce a functional adult walking leg.

Each functional walking leg is composed of an angular, flat *coxa* that lies close to the body, followed by a kneecaplike *trochanter*, a long *femur* and *tibia*, and finally a tarsus. The tarsus is usually subdivided into a series of five subsegments, the *tarsomeres*, and in some groups the tibia and several tarsomeres may have spines or spurs. The last tarsomere, the *distitarsus*, has a ventral padlike *arolium* flanked by two claws and two padlike *paronychia* (Clench 1975).

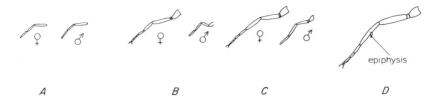

epiphysis

A   B   C   D

28. The forelegs of butterflies vary considerably and have evolved under different selective pressures. The nymphalid foreleg (A) is reduced to a three-segmented stub in both sexes. Those of riodinids (B) are reduced in the male to four segments, but virtually unchanged from the papilionid plan in the female. Likewise, female lycaenid forelegs are five-segmented but those of the male (C) are slightly reduced. The legs of papilionids (D) are believed to represent the ancestral form with all segments and spines well developed in both sexes. (After Clench, in Howe 1975.)

This basic leg plan is often modified, illustrating the evolutionary plasticity of the jointed arthropodan appendage. For example, the legs of the first thoracic segment or prothorax—which does not bear wings—are often reduced to stubs composed only of the coxa, trochanter, and tibia. These stubs are normally held tightly to the body so that some butterflies appear to have only two pairs of legs— one on the middle segment or mesothorax, and one on the third and last thoracic segment, the metathorax.

The forelegs are almost always smaller than the meso- and meta-thoracic legs, and in some female nymphalid butterflies at least, the ends of the forelegs are equipped with sensory pits and hairs. Some of these pits and hairs help females to identify suitable larval food plants (e.g., Chun and Schoonhoven 1973). For example, a female Red Admiral (*Vanessa atalanta*) locates potential larval food plants and typically "drums" the surface of the leaves rapidly with the stublike forelegs—perhaps to sample the roughness or texture of the plant, perhaps to abrade the surface and release plant compounds that indicate whether or not the plant is acceptable. No one knows precisely what such drumming action accomplishes, but it is likely a sensory action used to determine the suitability of the leaf for oviposition, as well as recognize a potential larval host plant (Calvert 1974).

In some families the morphology of the foretarsi differs between the sexes. Those of male nymphalid and lycaenid butterflies may be

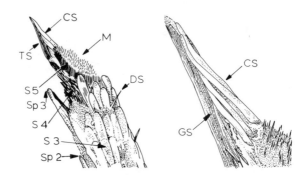

29. A scanning electron micrograph shows a lateral view of the Bordered Patch (*Chlosyne lacinia*) foretarsus. The foretarsus is composed of three subsegments (tarsal segments 3, 4, and 5) equipped with dorsalateral sensilla (DS) and a special arrangement of ventral spines (Sp) and their associated sensilla. A terminal spine (TS) and the clustered trichoid sensilla (CS) form the tip of the distitarsus (at 181×). (Microtrichia = M.) The tip of the distitarsus is magnified further (672×) in the right hand electronmicrograph. The bladelike spine with its groove (GS) is shown, as are parts of five trichoid sensilla (CS). (From Calvert 1974.)

fused together and covered with long hairs and scales while those of females are segmented. Using the scanning electron microscope, Calvert (1974) examined the microstructure of tarsal receptors in the females of the Bordered Patch (*Chlosyne lacinia*). Earlier, Fox (1966) hypothesized that females drum the leaf surface with their foretarsi in order to release "essential oils" that could be sensed by special *trichoid sensilla*—long, spearlike sensory structures that lead to circular sockets in the exoskeleton. And Myers (1969) determined that oviposition by females was inhibited when the foretarsi were removed.

Chlosyne lacinia females use the spines on the end of the foretarsi to tap the upper surface of the leaf. During this activity the antennae bob up and down toward and away from the leaf surface. If a female is satisfied with her choice, oviposition begins. If not, she pauses briefly and flies away. Calvert's electron micrographs show that scales extend only to the distitarsus, while immediately beyond this lie many microtrichia.

Each segment, except the distitarsus, also bears a pair of spines, whereas the trichoid sensilla are found only on the terminal and subterminal segments. The sensilla lie close together within a lateral groove of the spine and twist around each other to form a pointed, braided structure. Calvert envisions the spines abrading the leaf surface during drumming, thereby releasing plant juices to collect—perhaps by capillary action—in the groove of the spine. The clusters of trichoid sensilla are positioned exactly as would be required to detect the chemical properties of plant juices.

In addition to the muscles, the thorax also houses several "tubes": a hollow dorsal aorta, a solid ventral nerve cord, and a hollow esophagus that leads to a crop in the abdomen. The aorta is assisted in its periodic pumping action by *accessory pulsatile organs*, which force haemolymph into the vein system of the wings. There are other modifications, but basically the circulatory system of an adult is similar in overall design to that of the larva: both consist of a dorsal hollow tube with paired segmental holes or ostia that allow the haemolymph to percolate into the heart interior.

Wasserthal (1976) discovered that the direction of heart muscle contraction changes periodically, so that the haemolymph is sometimes pushed backward toward the abdomen. Abdominal expansion appears to assist this reversal. In fact, the abdomen produces volleylike ventilatory contractions during the forward heart pulse as well (Wasserthal 1976). Even resting butterflies have a regular heartbeat that is coordinated with the activity of the abdomen and the accessory pulsatile organs (Wasserthal 1974, 1980). He proposes that heartbeat reversal facilitates haemolymph exchange and air ventilation in the anterior body and wings. Interestingly, Wasserthal discovered the periodicity of heartbeat reversal by measuring temperature differences at different locations along the body.

It now appears that the heart, abdominal muscles, and the accessory pulsatile organs are all involved in the periodic and controlled transport of haemolymph. There is also a conspicuous ventral diaphragm in butterflies with an uninterrupted and uniform muscle activity that undulates independently of all other pulsatile systems, but it is not as effective in transporting haemolymph as insect physiologists had once believed, according to Wasserthal. However, the diaphragm probably aids the distribution of haemolymph throughout the abdominal cavity.

The pumping of the abdominal heart and the momentary contraction of the abdominal muscles also forces the haemolymph to flow

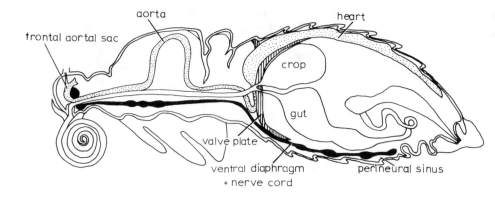

30. A schematic sagittal view of a butterfly body shows the distribution of
the organs involved in haemolymph circulation of the Old World Swallowtail (*Papilio machaon*). Note that the ventral diaphragm and nerve cord
are united to form a single organ that extends back toward the sixth
abdominal segment and channels the flow of haemolymph in the perineural
sinus. (From Wasserthal 1980.)

into the thorax. At the moment of contraction, if the abdomen stays
nearly the same length, the haemolymph pressure in the thorax will be
higher than that in the abdomen. At this point, air could run through the
abdominal spiracles into the tracheal system to balance the pressure
difference. As blood percolates back into the heart from below the
ventral diaphragm, the air would be expelled. Thus, the abdominal
muscles and heart are also indirectly involved in tracheal ventilation.

If the heartbeat is reversed so that the pulse is initiated anteriorly
and passes the haemolymph toward the abdomen, then air should
rush into the thoracic spiracles to balance the lower pressure within
the thorax. And if the direction of the heartbeat is reversed periodically, then both anterior and posterior areas in the body can
receive sufficient nutrients from the circulating haemolymph, as
well as enjoy increased ventilation. The slow change in abdominal
length involved in this oscillating haemolymph transport is a recent
discovery, the ramifications of which are only now being investigated (Wasserthal 1980).

The thorax is also the locomotor unit, crammed full of muscles to
operate the legs and especially the wings of the mesothorax and

metathorax. The muscles that operate the legs and wings are made up of striated fibers similar to the fibers composing vertebrate muscles of the limbs. Like those of other insects, the muscles of butterflies are very small, yet seem endowed with supernatural strength. This is because muscle power is proportional to cross-sectional area—the greater the cross-sectional area, the stronger the muscle appears relative to its size. As muscles become longer, however, the ratio of the cross section to length decreases, and the muscles become relatively weaker. If men were the size of butterflies, their muscles would be equally strong.

Mitochondria are the "power packs" of the muscle cells where chemical respiration and energy release take place. These tiny subcellular structures may make up over 20 percent of the total volume of the flight muscles. To respire under the strenuous conditions of flight, the muscle fibers are supplied directly with their own air supply from the *tracheoles*. These tubelike tracheoles then coalesce into larger trunks or tracheae that lead to the spiracles. The spiracles are paired openings that are found along the sides, or *pleura*, of the thoracic and abdominal segments.

The wings are the crowning glory of every butterfly. They are composed of a system of branching struts, the veins, containing tracheae, associated nerves, and circulating haemolymph. The branching veins are sandwiched between two transparent layers of chitin, a tough polymer composed of sugarlike units that have had nitrogen bonded to them. Chitin is flexible, resistant to chemical attack, yet rigid enough to be a protective and supporting layer of the butterfly exoskeleton.

The wing surface is divided into several regions that are useful in taxonomy. These regions are not always well defined, however, because the shapes of butterfly wings vary considerably between and within species. For example, the wings of the Zebra butterfly (*Heliconius charitonius*) are extremely long. Their narrowness permits unusual aerial acrobatics such as flying sideways, backward, and even upside down. In contrast, the wings of the Question Mark (*Polygonia interrogationis*) are blunt and triangular in shape, but jagged at the margins. And sometimes there are sexual differences: for example, the forewings of male Gulf Fritillary butterfly (*Agraulis vanillae*) are more pointed and narrower than those of the female.

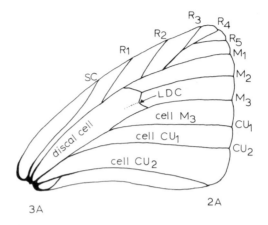

31. The veins and cells (interspaces) of a generalized butterfly forewing. SC = subcosta; R = radius; M = media; CU = cubitus; A = anal (or vannal); LDC = lower discocellular cross-vein. (After Clench, in Howe 1975.)

The major veins, in order from front to back, are the *costa*, the *subcosta*, the *radius*, the *media*, the *cubitus*, and one or more *anal* veins. The vein pattern is taxonomically important, and identification to genus can often be made on this basis alone. Areas confined by the veins are termed cells, and these likewise have characteristic shapes. The major interior cell of each wing is the *discal cell*.

While the wings of butterflies are *analogous* to (serve the same function as) the wings of bats and birds, they are derived from very different embryological tissues and operate very differently as well. Whereas the wings of bats and birds have internal muscles, those of butterflies have none at all. Instead, butterflies adjust their wings in flight with a series of direct muscles attached to the base of the wings. However, butterfly wings are powered primarily by indirect muscles—massive longitudinal bands attached to internal plates (phragmata) between the mesothorax and metathorax, and by even larger dorsal-ventral muscles that extend up and down within both winged segments.

The flight mechanism of butterflies is so complex that a book entirely on that subject would be required to do it justice. Basically though, the wings are inserted between the top, or terga, of the thorax, and the sides, or pleura, in such a way that the wings act like

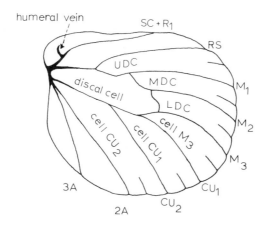

32. The veins and major cells (interspaces) of the hind wing of a generalized butterfly. The abbreviations in figure 31, and UDC = upper discocellular cross-vein; MDC = middle discocellular cross-vein; RS = radial sector. (After Clench, in Howe 1975.)

levers positioned on fulcra. During the upstroke, the dorsal-ventral muscles contract, forcing the edge of the terga down on the bases of the wings, while the pleura are forced outward and the distal ends of the wings upward. The downstroke requires the opposite actions: the dorsal-ventral muscles relax and the longitudinal muscles running between the segments contract, the tergum is forced up, the pleura are pushed inward toward the midline of the body, causing the wings to fall down.

Thus, the upstroke and downstroke are powered by the alternate contraction and relaxation of these opposing pairs of indirect flight muscles. However, these muscles really only deform the shape of the thorax in a rhythmic manner. It is the direct muscles that allow the wings to be twisted and inclined at different angles relative to the path of movement, thereby controlling the angle of attack. Controlled flight, then, requires both direct and indirect muscles operating in coordination.

Because there are two sets of indirect flight muscles, it seems possible that the wings of butterflies might operate independently, but they do not. Instead, both pairs function as a single aerodynamic unit, raised and lowered by the synchronous contractions of the

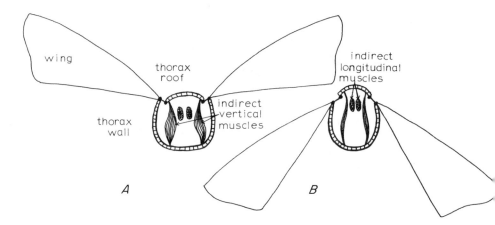

33. The lever-fulcrum action of the wings-thorax during butterfly flight are shown in this diagram. Wings are raised (A) as the dorsal-ventral muscles contract and the longitudinal muscles relax, thereby lowering the tergum of the thorax and forcing the wings up. When the longitudinal muscles contract and the dorsal-ventral muscles relax (B), the tergum is puckered and the wings are lowered. (After Dalton 1975.)

same muscles in both thoracic segments. And because the anal area of the forewing overlaps the costal margin of the hind wing during the downstroke, this synchronized activity of the indirect flight muscles is reinforced.

Although butterfly wings are comparatively flat, their anterior margins are thickened, giving them a slight curvature in flight similar to that of an airfoil. During flight, a butterfly moves its wings in a slanted figure eight, similar to the way a person would tread water with his arms. Controlling the angle of the wings determines whether the insect will propel itself forward or backward, or hover in the air—a metabolically expensive way to fly.

As air flows over the curved butterfly wing/airfoil, it must flow faster and farther over the dorsal surface than across the ventral surface. Flight is generated when the upward force or lift exceeds the downward-pulling force of gravity. For an average-size butterfly, the flight muscles are very strong relative to the weight of the body, and the wings are broad, flexible, and tough. As a result, enough lift is generated to make a butterfly airborne with but a few flaps of the wings.

34. The upstroke of the Monarch butterfly (*Danaus plexippus*) elegantly shows the wings as curved airfoils. (After Dalton 1975.)

The forward motion provided by the angled upstroke and down-stroke of the wings creates thrust, but wind exerts a force termed drag on any object. The greater the surface area presented to the wind, the greater the *form drag* of the object. Form drag is greatest when a butterfly's wings are caught vertical to a gust of wind. For this reason, butterflies are usually not seen in extremely windy, open spaces, and sometimes not at all, even on warm, but windy days. Lift also varies with the inclination of the wings, hence the angle of attack to the wind. The greater the angle of attack, the less lift. Some insects can lift themselves almost vertically off a perch like a helicopter, but all butterflies, save for perhaps the hovering heliconiines, have some nonvertical angle of attack during flight.

The pictures of the Monarch butterfly (*Danaus plexippus*) illustrate wing position and flexibility during flight. In the upstroke the forewings are brought slightly forward, twisting their anterior margins. Notice how the two wings overlap and form a continuous curved airfoil. In the downstroke the legs are often brought tightly against the body while the abdomen flares slightly upward.

The fastest butterfly (excluding the skippers, which are not true butterflies) rarely exceed twenty miles per hour. This means that even the stereotypical bespectacled, knock-kneed lepidopterist can

35. The power of a Monarch in flight is illustrated in this sketch of the butterfly half-way through the downstroke. (After Dalton 1975.)

catch them. Some butterflies, especially the larger danaids and papilionids can glide for several minutes on thermal uplifts. The gliding of the Monarch is an art form in itself, and the flight of the Zebra butterfly is precise enough to allow it to flip between the silk supporting lines of the huge orb-weaver spider, *Nephila clavipes*. Yet despite these interesting behaviors and the physical phenomena that must be their basis, many aspects of butterfly flight are virtually unexplored.

The wings of butterflies are completely clothed in scales. Other insects such as mosquitos and silverfish also have scales, but these evolved independently in each group. Furthermore, the scales of butterflies vary greatly in shape, color, and function, and are far more spectacular than those of other insects. Each scale has a pedicel or base mounted in a doughnut-shaped cell, the *tormogen*, on the wing surface. The scale forms in nearly the same way as the microtrichia that cover the body and basal regions of the wings. A huge epidermal cell, the *trichogen*, produces a club-shaped process that eventually flattens out into a scale. As the cell membrane of the scale breaks down, a complex internal gridwork is revealed, punctuated by pores that fill with air.

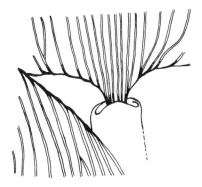

36. A scale in its tormogen socket—a drawing of an electronmicrograph. (From Downey and Allyn 1975.)

Every butterfly wing is covered with thousands of tormogens, each harboring the pedicel of a flattened scale. The scales overlap each other like shingles on a roof, but these "shingles" are only attached by the weak pedicel. Even the slightest touch will remove dozens of the dust-sized scales. Some claim that this characteristic enables butterflies to escape from the sticky strands of spider webs. The viscid strands may remove many of the loosely attached scales, but as a trade-off the butterfly escapes with its life.

Butterfly scales serve a variety of unexpected functions. For example, they increase the mass of the wings and therefore the wings' capacity to retain heat. And because scales are air filled, they also

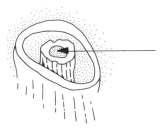

37. A scale of the Silvery Blue butterfly (*Glaucopsyche lygdamus*) is shown severed at its pedicel. Note the tubular nature of the pedicel and the tubular nature of the tormogen socket—a tube-within-a-tube arrangement that is so common in natural design. (After Downey and Allyn 1975.)

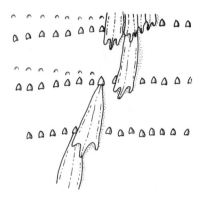

38. Drawing of a section of a *Morpho polyphemus* wing shows the linear arrangement of the tormogens, as well as the overlapping scales that form the shinglelike arrangement. (From Downey and Allyn 1975.)

can serve as effective insulators for the body. Furthermore, the complex internal gridwork of the scales absorbs solar radiation more efficiently than would a flat scale without internal struts. All these physical attributes, as we shall see, are important to temperature regulation (chap. 4).

Scales are also important in generating thrust. Butterflies with

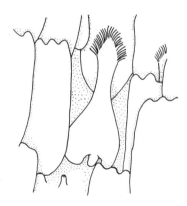

39. The odd-shaped "battledore scales" of the Cabbage butterfly (*Pieris rapae*) are shown here in this drawing from an electronmicrograph. The battledore scales are believed to contain or distribute pheromones used in sexual courtship. (From Downey and Allyn 1975.)

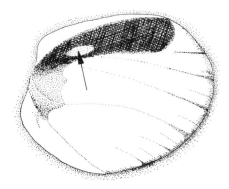

40. The sex patch of the male Dainty Sulphur (*Nathalis iole*) shows its location within a band of melanic scales on the leading dorsal edge of the hind wing. (From Vetter and Rutowski 1978.)

their scales removed lose about 15 percent of their effective thrust during flight. Of course, a large butterfly like a Monarch may bear over five hundred thousand scales on its wings alone, so there are plenty of scales to lose before flight is affected.

Finally, some scales like the *battledore*, or "tennis-racket" scales of male lycaenid butterflies (Downey and Allyn 1975) are responsible for the production of sexual scents (pheromones) that are important in mating rituals (see chap. 7). These *androconial scales* are similar to those found in the *sex patches* or *sex brands* of other butterflies such as male Dainty Sulphur butterflies (*Nathalis iole*). The male androconia have a small gland at their base that produces the pheromone. The fluid is drawn up by capillary action through the hollow pedicel into the internal structure of the scale, and diffuses to the minute hairs on the exterior scale surface. These fine hairs increase the surface area available for evaporation, thereby making the odor seem stronger because more of it diffuses per unit time (Rutowski 1979) than would from a blunt-edged scale.

Other butterflies like the male Monarch have evolved eversible abdominal brushes that resemble the minute sails of milkweed seeds. During courtship the brushes are everted and inserted into large sex patches on the hindwings bearing androconia. With every wingbeat, the female Monarch is fanned with enticing pheromone (Pliske and Salpeter 1971).

Recent experiments with the Little Yellow (*Eurema lisa*) and the Common Sulphur (*Colias philodice*) indicate that both scent patches and special scent scales scattered over the wing are involved in the dissemination of pheromones. The Little Yellow has a scent patch and androconial scales in the *friction area* of the ventral forewing where the forewing overlaps the dorsal surface of the hind wing. In contrast, the friction areas of the Common Sulphur are located on the dorsal hind wing where the hind wing overlaps the ventral surface of the forewing. These scent patches produce specific chemical signals designed to "seduce and settle" responsive females during courtship. Responsive females extend the abdomen ventrally between the closed wings, a behavioral posture that permits copulation (Vetter and Rutowski 1978).

Other sulphur butterflies (Coliadini) have scent scales or scent patches on different areas of the wings. The oval sex brands of the Dainty Sulphur are located in the friction area of the dorsal hind wing where they are overlapped by the ventral forewing. These sex brands change color rapidly with age or after death from a bright orange to a dull yellow (Clench 1976a). The sex patch in this species is outlined in melanic scales, making it clearly visible, but the interior of the patch may be devoid of scales, or covered with scales that are shriveled in appearance (Vetter and Rutowski 1978).

The orange, triangular sex brands on the ventral forewings of the Sleepy Orange (*Eurema nicippe*) are always covered by short scales at twice the density of an average wing surface. Once again, though, the sex patch is located in the friction area where forewings and hind wings overlap, perhaps, as Rutowski (1979) hypothesizes, to reduce evaporation. The location of the sex patches in the Coliadini contrasts sharply with the abdominal sites of the eversible "hair pencil" brushes found in the danaids (Pliske and Salpeter 1971). Yet, the similarities between androconial scales in general illustrate either a common evolutionary origin, or the action of similar selective pressures (Vetter and Rutowski 1978).

The wing patterns of butterflies are actually mosaics constructed from many different kinds and colors of scales. Each scale, however, is of only one color. But color is really the brain's interpretation of different wavelengths of light, and there are several ways in which scales can exhibit colors. For example, scales may be dyed, quite literally, during their formation, with organic pigments.

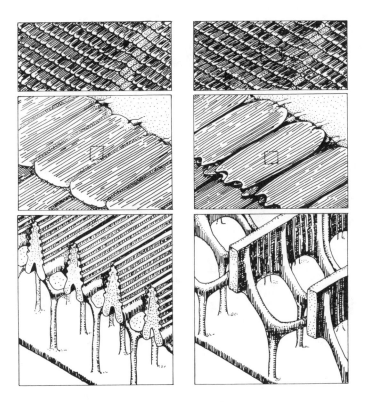

41. Iridescent scales of Morpho butterflies usually have a dark melanic pigment that absorbs most visible transmitted light and reflects wavelengths such as blue and green. The iridescence is caused when these wavelengths are reflected on the thin vertical vanes on the scale surface, or by horizontal stacks of thin lamellae making up the outer cuticle of the scale. The combination of mechanically reflected light and light reflected by pigments makes the iridescence especially brilliant. (From Nijhout 1981.)

A pigmented scale emits a particular color because it absorbs all wavelengths of light except the ones we perceive. Visible wavelengths are thus reflected, not absorbed, by scale pigments. Some scales appear black because they absorb all wavelengths of light, at least in the visible region to which our eyes and brain are sensitive. Likewise, a red scale appears red because it reflects wavelengths predominantly in the red wavelengths of the visible spectrum while absorbing all others.

Melanins are largely responsible for the blacks, grays, browns, tans, rusts, reds, and yellowish colors, while pterins produce the brilliant reds, oranges, yellows, and some white pigments. Other organic dyes—the flavones, carotenoids, and ommochromes—account for colors ranging from ivory to dark yellow (Wigglesworth 1972; Clench 1975; Smart 1975). Other scale colors such as some whites and all true iridescent blues and greens are caused by structural colors. For example, a scale may appear white because all light incident on the scale is reflected from tiny air bubbles within the structural gridwork of the scale. Likewise, the iridescent blue scales of the tropical genus *Morpho* result from the reflection and reinforcement of blue wavelengths of light from ridges located on long vanes that run the length of the scale. These ridges are about 0.22 micrometer apart, or approximately half the wavelength of blue light. When the sunlight strikes the ridges perpendicularly, the blue wavelengths are reflected and reinforced over a narrow wave band of blue, while other wavelengths are absorbed by pigments such as melanins within the cuticular ridges (Nijhout 1981).

The wavelength of blue light emitted depends on the viewing angle and the light's angle of incidence. When the viewing angle changes, the color changes appreciably. Usually, the intense metallic blues are visible over a relatively narrow angle and only if the angle of light incidence remains unchanged. If the scales are immersed in a solvent, the cuticle absorbs the liquid, expanding the distance between the *microstriations* or ridges. As a result, the scales reflect a different wavelength, giving them a metallic iridescent color such as green. Silberglied (1979) has shown that *Morpho* scales are even more spectacularly reflective in the ultraviolet than they are in the blue.

If pigmented scales are intermixed with or overlapped by structural scales, the effect can be awesome, producing the most complex and intense color mixtures in the animal kingdom, although tropical bird and fish lovers undoubtedly would give me a strong argument here. The difference in the micromorphology of iridescent and pigment scales is depicted by the fine structure of the *Morpho rhetenor* scales in figure 41. The noniridescent scales usually lack the horizontal microstriations, and so their vanes are blunt-ended, not pyramidal in shape. However, some shiny transparent areas such as eyespots contain neither pigmented nor structural scales. Instead,

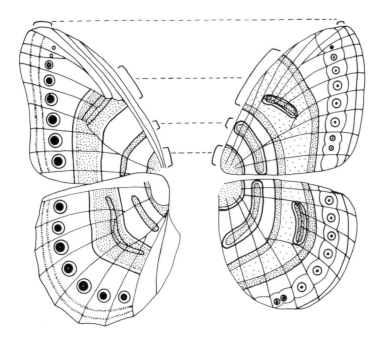

42. The "nymphalid" ground plan can be used to derive most of the color patterns found on butterfly wings. (From Nijhout 1981.)

their transparency is due to the lack of scales and the transparency of the cuticle exposed below.

As a rule, a single butterfly has only four or five different pigments in its scales. But the scales are so small—only 100 micrometers by 50 micrometers—that thousands of differently colored scales may be optically blended by our brains to produce a myriad of different hues and intensities. In some butterflies, for example, the ventral surface of the hind wings appears green (e.g., *Euchloe*) because of the juxtaposition of black and yellow scales. Mixing colors to produce different effects apparently is a more ancient trick than modern artists might care to believe.

The physiology of pigmented and structural colors is fairly straightforward compared with the incredible diversity and complexity of butterfly wing patterns. In the 1920s two European researchers, Schwanwitsch (1924) and Süffert (1927), independently determined that most lepidopteran wing patterns could be derived from a basic *nymphalid plan*. By altering the form, thickness, and color of the lines and eyespots of this hypothetical plan, one can derive nearly any butterfly wing pattern.

This hypothetical wing pattern is not necessarily the ancestral wing pattern, but simply a model that best explains the diversity of lepidopteran wing patterns. It can be used to detect and describe homologous (those of common origin) color patterns of different species. This diversity of wing pattern must be genetically based since the wing patterns are species-specific. Each wing begins in the pupa as an unpigmented double layer of epidermal cells. The genes involved in forming the microstructures of the wings, and those that regulate the biochemical pathways in which pigments are synthesized and structural scales are constructed, are responsible for the final color and pattern of the wings during development (Nijhout 1981).

Nijhout (1978) views the wing patterns of butterflies as complex mosaics of cells. He has determined that the pattern of each wing cell (between veins) is expressed independently of the patterns in other cells. Furthermore, pigments are deposited in the scales in a specific relationship to a group of organizing epidermal cells or "focus." A focus always lies in the cell midline and commonly appears as a minute pigmented spot. The simplest pattern is a circle or series of concentric circles (eyespots) in which the focus serves as a reference point. It is here that positional information (for pigment deposition) is specified. A circle or eyespot results when all points equidistant from a focus undergo identical differentiation (Nijhout 1978).

The complex and highly variable wing patterns consist of a relatively small number of elemental patterns. Süffert (1927) recognized five basic types of patterns: ripple or striped patterns running perpendicular to the wings; dependent patterns whose position depends on the topographical features of the wing; crossbands; eyespots or ocelli; and color fields that involve large patches of color. One of the most complex patterns is a type of crossband system known as the central symmetry system. It lies in the middle of the wings with an axis of symmetry passing through the discal spot, located at the apex of the discal cell and nearly equidistant from the base and distal margin of the wing (Nijhout 1978).

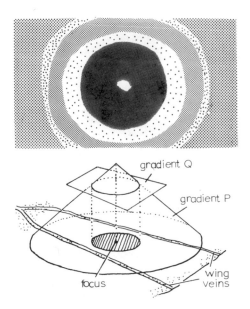

gradient Q

gradient P

focus

wing veins

43. Circular border ocelli can be diagrammatically produced by assuming that two gradients, P and Q, act as a plane transecting a cone. If a chemical P diffuses equally in all directions from the focus of a wing cell, a cone is produced with the highest concentration of P at the cone's apex. The gradient represented by Q may be linear or nonlinear. (Fron Nijhout 1981.)

It was this and other symmetry systems, each confined to fairly well defined wing areas, that led to the development of the prototype nymphalid wing pattern proposed by Schwanwitsch and Süffert. Pigments that make up these symmetry systems are deposited with incredible accuracy in precise spatial relationships to one or more foci in each wing cell. According to Nijhout (1978), each species' color pattern is determined by the number and position of these foci on the wing surface, and by certain rules that specify the spatial relationships of these pigments to their foci.

The bottom line is that the entire butterfly "ground plan" develops according to these discrete organizing sources, the foci. Each focus controls the development of pattern within one wing cell only, or in an adjacent cell, if neighboring foci are lacking. Finally, species-specific patterns and variations on the nymphalid ground plan arise because of differences in the location of foci and dif-

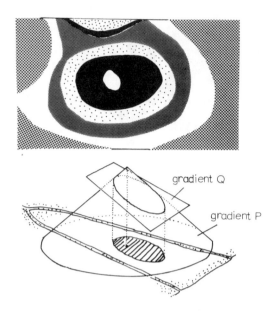

44. Elliptical ocelli can be generated by plane Q transecting cone P at an angle to the wing surface. (From Nijhout 1981.)

ferences in the rules that determine the shape of the pattern that develops around a focus. Any shifting of these foci—toward the base or margin of a cell, for example—can alter the wing pattern.

So far the presumed existence of developmental foci offers the simplest explanation for a number of experimental and morphological findings on the color patterns of Lepidoptera. For example, cautery of a focus on the dorsal side of a pupal wing does not affect an eyespot in the same position on the ventral side. Thus, foci of the dorsal and ventral wing surfaces are independent and separable (Nijhout 1980a). Nijhout (1980b) showed that the development of the large eyespot on the forewings of the Buckeye butterfly (*Precis coenia*) depends on the presence of a small focus of cells (probably fewer than three hundred) that lie on or near the center of the wings. The determination of the eyespot appears to be an intrinsic property of the focus, not dependent on its environment (Nijhout 1978), and eyespots can be transplanted from an area where they would normally occur to one where they would not.

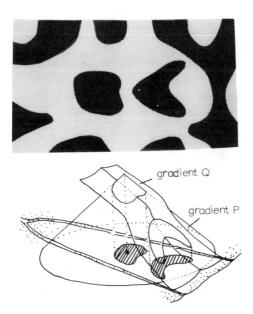

gradient Q

gradient P

45. Complex ellipses or crescents with one bifurcated end can be generated by assuming that plane P transects cone Q at a sharp angle. The actual pattern on the wing arises from the projection of the conic section on the wing. (From Nijhout 1981.)

According to Nijhout (1980a, 1980b) the response of the eyespot to cautery or transplantation is most readily interpreted if we assume that the focus is the source of a morphogen that is somehow able to induce the synthesis of specific pigments. In the Buckeye, all pigments are due to melanins, and thus it is necessary only that the morphogen control the differential synthesis of four enzymes (tyrosinases), each specific for the synthesis of one color.

Morphogen concentration is highest in the vicinity of the eyespot focus and hence no pigment is synthesized here. In areas where the concentration of morphogen is high, but below a critical level, only black (melanin) is produced, forming the central body of the eyespot. Still lower concentrations yield the synthesis of a buff pigment, and the lowest pigments produce the brown peripheral ring of the eyespot. Nijhout (1980a) observes that the presence of sharp boundaries between any two pigmented areas illustrates the accuracy with which the scale cells can measure the morphogen gradient.

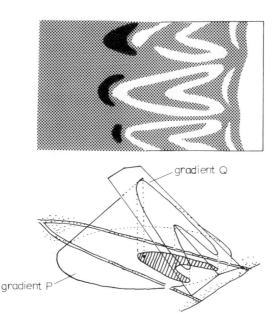

46. Even more complex patterns such as interrupted ellipses with bifurcated lobes or wavy marginal lines are possible. Again, it is the projection of the conic section that would give the pattern to the wing surface. (From Nijhout 1981.)

True circular patterns on butterfly wings are not as common as crescents, lines, and complexes of wavy lines such as those of the prominent central symmetry band system. Noncircular wing patterns such as parabolic designs and ovals usually appear as bilaterally symmetrical patterns whose axis of symmetry extends through a focus down the middle of a wing cell and parallel to the veins. This implies that the pattern "signal" from the presumed developmental focus does not necessarily travel in all directions at the same rate and intensity, or that all cells receiving the signal do not respond to the same degree.

Nijhout hypothesizes a signal in the form of a chemical substance P that is synthesized at a developmental pattern focus and then diffuses outwardly and equally in all directions. The resulting concentration gradient could be visualized as taking the shape of a cone in which the concentration of P drops in a mathematical way in all

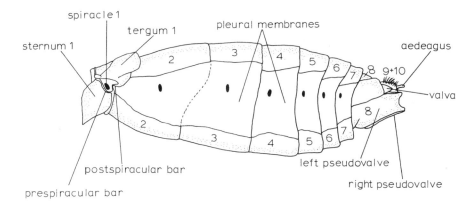

47. This diagram of the male abdomen of the Monarch butterfly (*Danaus plexippus*) shows the sclerite arrangement, the placement of spiracles, and the location of the external genitalia. (After Ehrlich and Ehrlich 1961.)

directions away from the focus. If scale cells are genetically programmed to synthesize, say melanin, whenever P reaches a certain concentration, then the result is a black ring, just as if a plane Q had transected the cone-shaped gradient of P a given distance from the top of the cone and then had been imprinted on the surface of the wing. In this scenario, circles and nonlinear patterns are formed when the genes of epidermal cells are stimulated to produce certain pigments, or stimulated to deposit organic compounds available to them in the wing cuticle.

In the diagram, the height at which plane Q transects the cone corresponds to the critical concentration of substance P required for the synthesis and/or deposition of melanin. Each pigment is thus represented by different planes transecting the cone of P at different heights and angles, that is, different ratios of P and Q. The width of the pigmented ring would be proportional to the thickness of plane Q transecting the cone.

If responsive cells at certain levels of P are not evenly distributed around a focal point, the result would be a nonlinear, but symmetrical, design, much as if our hypothetical cone of P had been cut at different angles with our pigment plane. Since the pattern on a butterfly's wing is genetically determined, perhaps the butterfly wing is like a silk screen with the design genetically etched upon it,

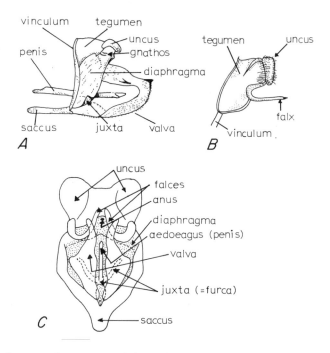

48. The internal structures of the male genitalia are shown for a generalized butterfly (A) and the dorsal structures of a lycaenid (B). The position of the internal structures of the male lycaenid (C) are shown in this ventral view. (After Klots, in Howe 1975.)

but with the genes of each cell awaiting their critical concentrations of one or more chemicals produced by the developmental foci.

We must leave the wings and their mysteries now, and devote some attention to the abdomen, the vital repository of most of the butterfly organs involved in digestion, excretion, and reproduction. Compared to the complexity of the head and thorax, the anatomy of the abdomen is relatively simple. Superficially, the abdomen is cigar-shaped: wider in the middle and tapering gently at the posterior and anterior. Each of the eight pregenital segments consists of a sclerotized dorsal *tergite* and a ventral *sternite*, separated by an unsclerotized and slightly more flexible *pleurite* on each side. Each pleurite bears a pair of spiracles, one on each side of the body. The abdomen also houses the remainder of the ventral nerve cord, the dorsal heart, and the malpighian tubules. Most importantly, the last

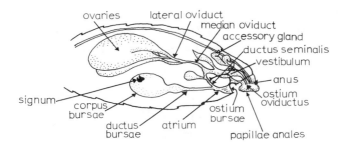

49. The internal anatomy of a generalized female's genitalia are as complex as those of the male. (After Klots, in Howe 1975.)

two segments bear the genitalia, which are remarkably dimorphic between the sexes. The male genitalia, in particular, are of special taxonomic importance.

A female is inseminated when the paired *valvae* or *claspers* of the male grasp the genital region of the female, while a forked structure, the *furca*, guides the penis through the female opening or *atrium* and into the *ductus bursae*. The heavily sclerotized penis passes a sperm package, the *spermatophore*, through the ductus bursae and into the corpus bursae for storage. The teeth within the corpus bursae rupture the sac, releasing the sperm to swim up the *ductus seminalis*. Here a sperm penetrates the micropyle of the egg as the egg passes down the *oviduct*.

Some lycaenids and especially primitive male papilionids—the alpine and arctic *Parnassius* in particular—secrete a peculiar structure around the female genitalia during mating, the *sphragis*. The sphragis may prevent other males from mating with the female, and ensures that only the sperm from the first male is allowed to fertilize the eggs. The shape of the sphragis is species-specific and of taxonomic importance (see also chap. 7).

Although it is not within the scope of this book to exhaust all that is known about the anatomy and physiology of butterflies, it should be clear that the adult is far more complex than any previous stage of metamorphosis. However, there are many interesting phenomena to be examined at all life stages. A single chapter in a popular book cannot possibly describe even briefly all that is known or that needs to be known.

Unfortunately, many lepidopterists are concerned with collecting new records and naming new variations, but pay little attention to these fascinating phenomena. The truth of the matter is that there is so much to know and understand that a person could study a single species for the rest of his life and still not completely understand the physical, physiological, and behavioral phenomena that govern a butterfly's life history. The next five chapters will examine but a few of these fascinating phenomena, drawn largely from the life histories of American butterflies.

CHAPTER 4

Aspects of Survival

Although flowers provide a common if not major source of food for butterflies, not all adults feed on *nectar*—a solution of sugars, water, and occasional amino acids. These same essential dietary constituents can also be obtained from other sources. For example, the adults of many species obtain part or all of their nutrient requirements from a variety of resources, including rotting fruit and vegetable matter, carrion, dung, perspiration, urine, pollen, grass inflorescences, and virtually any sweet secretion—whether it be from a flower, a hummingbird feeder, or special secretory glands of other insects. There are even ithomiid "antbutterflies"—butterflies that follow army ants to feed on antbird droppings (Ray and Andrews 1980).

As a group, in fact, adults are far less choosy about their liquid meals than their larvae are about their leafy meals. Larvae, of course, have little choice in the matter, because adult females chose the larval food plant in most cases. As with many aspects of butterfly ecology, the relationships between adult butterflies and their food resources have only recently been systematically investigated, and what we know at present is based on only a few general and comparative studies. Anecdotal reports suggest that butterflies spend up to 50 percent of their flying time foraging for food. However, the amount of time spent foraging depends on many factors such as time of day, local climatic conditions, the type and abundance of resources present, and even the sex and physical age of the butterfly.

For example, Hackberry butterflies (*Asterocampa* spp.) rarely if ever visit flowers. When they do, they tend to visit flowers that may serve as a source of nitrogen (Neck 1983). This is especially true of gravid females that may not require the carbohydrate-rich nectar of most other flowers. In another study by Stanton (1984), searching constraints were found to influence foraging patterns in three species of *Colias* butterflies. She determined that the activity of egg-laying females is partitioned into periods of searching for larval food plants and periods of visiting flowers for nectar. Based on these and other studies, it is apparent that many different factors can affect foraging behavior.

As with most animals, sugars are the most common form of energy storage molecule respired by butterflies. Nonetheless, sugars are bulkier to store than fats, yield only about half the energy of fat per gram upon oxidation, and as a result, such ephemeral molecules of energy must be regularly replaced. Migratory butterflies such as the Monarch (*Danaus plexippus*) have solved this problem by converting carbohydrates to fats. This biochemical solution reduces their *wing load* considerably during their months-long annual migration.

There are few long-term records of flower visitations by butterflies, but the best "butterfly flowers" tend to be found in certain plant families. These include the milkweeds (Asclepiadaceae), the daisies (Compositae), the mustards (Cruciferae), the mints (Lamiaceae), the peas (Leguminosae), and the vervains (Verbenaceae). The plant family most commonly used, according to a study by Shields (1972), is the Compositae, with twenty-nine genera visited by adult butterflies.

Faegri and van Der Pijl (1971) characterize the flowers most frequently visited by butterflies as diurnal, brightly colored blossoms with a mild odor. The blossom rim is usually narrow, flat, and has large, blunt petals. Nectar guides—pigmented stripes on petals typically visible only in the ultraviolet wavelengths—are sometimes present to steer butterflies and other pollinators toward the repository of pollen and nectar. Certain species of butterflies habitually visit particular species of flowers, and in the process of probing for nectar (and sometimes pollen) it is inevitable that pollen sacs, or *pollenia*, will be transferred first to the pollinator and then from one flower to the next. This is often because butterflies will walk around the flower as they feed, and pollenia will stick to the proboscis, hairs of the head, spines on the tibia, and even the tarsal claws.

Thus, a great deal of cross-pollination may take place, even over relatively great distances, helping to "outcross" and homogenize the genetic pool of the species over a broader geographic range than might otherwise be possible with, say, plants that are wind-pollinated. If cross-pollination is common, then coevolutionary relationships may exist between some flower groups, for example the composites, and certain genera of butterflies. However, the degree of these relationships (if they exist at all) is unknown. In fact, Wiklund and his colleagues (Wiklund, Erikson, and Lundberg 1979) have suggested that butterflies such as the Wood White (*Leptidea sinapsis*) often carry little pollen, and are therefore actually parasitizing the visiting flowers. In a similar vein, Murphy (1984) found that some flowers visited by the Checkerspot butterfly (*Euphydryas editha*) had lots of pollen carried off, whereas others did not.

When nectar-bearing flowers are scarce, or when competition for nectar with other pollinators (such as bees) is high, butterflies may use some rather unorthodox sources for food. Neck (1980) reports that grass inflorescences may occasionally be used by adults at times when other nutrient resources are scarce or absent. In October, he found individuals of the Queen butterfly (*Danaus gilippus strigosus*), the Buckeye (*Precis coenia*), and the Painted Lady (*Vanessa cardui*) probing the bases of grass inflorescences in Texas. Apparently, some substance was being removed from several grass species, which included bluestem. Of course, it is equally possible that these species were merely attracted to some volatile, water-insoluble compound secreted by the plant's epidermal cells, and otherwise not gaining any nutritional substance.

While feeding on grass inflorescenses is unusual, some species of butterflies feed exclusively or nearly so on other nonfloral food sources. The Pearly Eye (*Lethe portlandia*) is a woodland species well known for its seeming "addiction" to mammalian excrement. The cryptically colored undersides of the adults make them almost impossible to see while feeding, if it can euphemistically be called that. Dozens of these *coprophilous* butterflies may fly up suddenly in front of a hiker on a woodland trail—ample warning of unpleasantness in the path ahead. Besides feces, these and related species of butterflies imbibe exuded tree sap, fermented fruits, urine, sweat, and carrion—a virtual smorgasbord of odious delights.

Although fats and carbohydrates may be available to butterflies that forage on nonfloral food resources, until recently it was a mystery why males of so many species were attracted to mud puddles, urine, and dung deposits. Arms, Feeny, and Lederhouse (1974) showed that, at least for *mud-puddling behavior*, the butterflies may be searching for sodium because this ion indicates where compounds containing nitrogen are present.

They also established that amino acids obtained from wet soils are incorporated into body proteins. Adler (1982) documents mud-puddling in the moths, so apparently mud-puddling is a common lepidopteran behavior. However, mud-puddling is largely a male-oriented behavior. Females typically make up less than 5 percent of the butterflies imbibing on wet soil (Downes 1973). Later, Adler and Pearson (1982) found that total sodium levels differed significantly in male and female Cabbage White butterflies (*Pieris rapae*). In some cases, butterflies (e.g., lycaenids) have been observed to discharge a liquid from the proboscis while puddling (Jobe 1977), but the reason for this "pumping action" (Reinthal 1963) is unknown. Perhaps it allows greater uptake of nitrogenous compounds.

Why is nitrogen so important? A reliable nitrogen supply seems to be required for cellular maintenance as well as for reproduction. In fact, some butterflies in the Neotropics, especially female helico-niines, consume both nectar and pollen. Pollen contains nitrogen in the form of free amino acids, and these free amino acids are required to produce the egg protein (Gilbert 1972).

Heliconius prefer pollen from *Crocea, Gurania,* and *Anguria,* and these plants are believed to be critical for the long life span and long, continuous patterns of reproduction in many *Heliconius* species (Boggs, Smiley, and Gilbert 1981). As we shall see, there is a close relationship between adult movements, the formation of roosting colonies, and local blooming plants (Ehrlich and Gilbert 1973; see also chap. 8).

In temperate North America, Watt, Hoch, and Mills (1974) showed that *Colias* butterflies prefer flowers with a low concentration of monosaccharides (simple sugars) and nitrogen-rich free amino acids. This mixture may be important for nutritional value as well as to conform to the biophysical constraints placed on the proboscis for efficient pumping (Kingsolver and Daniel 1979; see also chap. 3).

So, the search for sodium—that is, the search for a source of nitrogen—also explains in part the bizarre mud-puddling behavior in which hundreds of butterflies may jostle each other for room on the damp mud surrounding rain puddles. Presumably, rainwater leaches out mineral salts in surrounding soils and concentrates them in the mud puddles. By contrast, the edges of small, rocky streams are not commonly used, because leached minerals from surrounding soils have little opportunity to concentrate in marginal areas bordering the stream (Craw 1975; Adler 1982).

Many of the nymphalids and satyrids attracted to mud flats along slow-moving rivers are also attracted to rotting fruit baits. In Africa, both sexes of the Charaxidae are attracted to tree sap and fermenting fruit, but only the males are attracted to dung and carrion (Sevastopulo 1974). Several have suggested that sodium salts may be required more by the neuromuscular system of males because they must expend a great deal of energy actively searching for mates and therefore do most of the flying (e.g., Craw 1975).

Gilbert and Singer (1975) propose that males and females of some species partition nitrogen resources. In this scenario, females obtain their nitrogen requirements from flowers while males obtain theirs from many different sources including mud puddles, urine, carrion, and feces. Thus, females are spared intraspecific competition with males for scarce resources. They have shown that some female *Heliconius* butterflies fly earlier in the day than males and thereby harvest the most pollen and the best nitrogen supply. However, the males of some species (e.g., danaid butterflies) forage at some plants not to obtain nitrogen, but for specific compounds that serve as pheromones (scents used for communication) or precursors of pheromones (Edgar and Culvenor 1974).

On hot summer days, hundreds of butterflies—sometimes belonging to a single species—may completely cover a dung pile or urine patch. The butterflies may engorge themselves to the point where they periodically excrete drops of liquid. The search for sodium is so strong that many nymphalids (e.g., *Polygonia, Vanessa, Asterocampa*) will alight on a person's head or body to lap up the sweaty by-products of summertime activity. Such butterflies appear "tame," and they can be prodded gently into crawling onto an open finger and then lifted to view only a few inches from the eye. Here they remain, eye to eye, still sampling the skin with the proboscis. This unusual behavior has generated all sorts of anthropomorphic flights of fancy concerning communication between life forms at a cosmic level. True believers of "interspecific communication" would rather not hear the reductionist explanation: probing for sodium sounds oh, so mechanical!

Closely related butterflies usually choose similar adult foods. Perhaps this is simply an evolutionary legacy, but as Gilbert and Singer (1975) point out, dung, fruit, and sap eaters are generally *cryptic*—and edible—whereas many flower-visiting species are either small and difficult to capture, or distasteful and brightly marked with warning colors such as orange and black. The relationship between certain groups of butterflies and certain food resources is still poorly understood (see chap. 8) and a great many questions remain unanswered. For example, why do many species choose to nectar at and thus pollinate groups of flowering plants not related to their larval food plants?

Regardless of the source of nourishment, all foraging behavior is ultimately aimed at maintaining the butterfly body and fueling it for flight. Flight, of course, is a prerequisite for mating, oviposition, and dispersal. However, butterfly flight time periods vary considerably. Most species are diurnally active, but some Neotropical owl butterflies (*Caligo* spp.) are *crepuscular* and fly during twilight. Occasionally, even some temperate species such as the Question Mark (*Polygonia interrogationis*), the Red Admiral (*Vanessa atalanta*), and the Monarch are crepuscular. On hot summer nights, these species can be seen flying long after sunset, and they have even been reported to circle street lights. Typically, though, butterflies are creatures of the day, and it is unusual climatic conditions or a disturbance that induces them to forage at night. Neck (1976c), for example, reports the nocturnal activity of Monarch butterflies during an extended drought.

On the whole, butterflies are warm weather animals with a "cold-blooded" or *poikilothermic* physiology. This means that their rate of internal heat production is low relative to their rate of heat loss from the body surface. In the absence of sunlight, most butterflies cannot maintain a high body temperature. By contrast, *homeothermic* or "warm-blooded" animals such as mammals and birds usually have a continuously high rate of internal heat production, and can maintain a high body temperature with or without sunlight.

All butterflies need a high body temperature to fly well. A warm thorax is especially important because the muscles operating the wings and legs are located in the thorax. In fact, the thoracic temperature range of butterflies flying in the field is usually between 25 and 44°C (75 and 112°F), a temperature range remarkably close to that of homeothermic animals (Douglas 1979). Of course, coordinated flight is critical to the survival of adult butterflies because it allows them to escape predators and compete with other insects for limited nectar resources. Controlled flight also enables males to search and compete for available females, thus ensuring their reproductive success (Douglas 1979).

The Mourning Cloak (*Nymphalis antiopa*) has evolved both behavioral and physiological mechanisms by which it raises its thoracic temperatures 8 to 11°C (15 to 20°F) above the air temperature. So effective are these mechanisms that Mourning Cloak butterflies can even be seen flying on relatively warm days (13°C[55°F]) during the occasional winter thaws of January and February. Their ability to control the thoracic temperature over wide fluctuations in the thermal environment is termed *thermoregulation*, or temperature regulation. Thermoregulatory mechanisms are valuable adaptations particularly for those species that inhabit higher latitudes and elevations.

Among cold-adapted butterflies, the Mourning Cloak possesses yet another unusual method of surviving cold periods: *hibernation*. After emerging from the chrysalis in the fall, the adult butterflies feed on nectar and exuded tree sap, converting these carbohydrates into internal fat reserves. However, fall butterflies soon enter diapause—they do not mate and the sexual organs remain in an immature state. When air temperatures drop below 10°C (50°F) and day lengths become short, they search for safe dry crannies in log piles, fallen trees, or brick walls. In these protected and well-insulated places the adults avoid the vagaries of winter (Douglas 1978).

However, these protected niches can cause problems in the spring because sunlight has a difficult time penetrating down to the hibernation quarters. This problem is solved by a "shivering" mechanism appropriately termed *muscular thermogenesis*. During muscular thermogenesis, the opposing pairs of thoracic muscles are contracted and relaxed synchronously rather than alternately. Thus, the thoracic muscles do not move the wings in the wide arcs required for flight, but rather vibrate the wings in small amplitudes and produce a visible shivering effect. This shivering action produces an audible clicking sound from the wings, but more importantly, the intense muscular activity raises the thoracic temperature by 8 to 11°C (15 to 20°F) in a few minutes.

As a result, thoracic temperature may be boosted to 21°C (70°F) or more, high enough to operate the leg muscles efficiently and permit movement from the tight hibernating quarters to open sunlit areas. Of course, the normal alternating contraction of the opposing pairs of flight muscles produces an equivalent amount of heat as shivering, but full wing contraction would be nearly impossible in the confined hibernating sites. In addition, the butterfly cannot fly well at low thoracic temperatures, and any helpless flopping would needlessly expose the butterfly to hungry birds.

Even with these behavioral precautions, I have occasionally seen Mourning Cloaks attacked by birds during a metabolic warm-up. Under such life-threatening circumstances, the butterfly's defense is to "play dead." In this motionless state a Mourning Cloak can be pecked and thrown high into the air, but will drop to the ground without flexing a muscle, appearing to the predator as a lifeless wood chip or a fallen leaf—both objects that it strongly resembles with the wings folded. When confronted with such bizarre behavior, many avian predators stop the attack and leave behind what might have been a juicy morsel of protein!

Once a sunny and protected spot is located, the Mourning Cloak ceases shivering and like other butterflies raises its thoracic temperature even further by basking. Depending on the species, butterflies use one or more characteristic basking orientations. These include *dorsal basking* with the wings held completely open to sunlight, *body basking* with the wings angled so that only the body receives solar radiation, and *lateral basking* in which the wings are held closed over the body and the butterfly tilts sideways, thus presenting only the undersides of the wings to the sun.

Under most environmental conditions, different amounts of solar radiation are absorbed with each orientation; however, for most of the larger species the dorsal basking position is the most effective for rapid heat gain. In this position, radiant energy is absorbed most efficiently by the dark thorax and the bases of the wings adjacent to the thorax. The darkly pigmented wing bases with their relatively large surface area heat much more rapidly than the thorax and can reach a temperature of 45°C (115°F) in sixty seconds under a noonday sun.

In turn, some of the absorbed solar energy from the basal areas of the wings is conducted and radiated to the cooler thorax. However, the color and distribution of the wings' veins outside the basal area has no bearing on this thermoregulatory capacity. Butterflies fine-tune their body temperature by controlling the angle of the wings to the sun, a behavioral strategy that assures the proper thoracic temperature necessary for controlled flight.

In addition to capturing solar radiation directly, a butterfly's outspread wings allow the buildup of an insulating layer of air beneath the wings, thus buffering the thorax and protecting it from cooling breezes. The thick and uniformly hairy wing bases that envelop the butterfly's body during basking also reduce heat loss by increasing the effective diameter of the thorax. Of course, larger butterflies lose energy more slowly than smaller butterflies of the same temperature because the surface-area-to-volume ratio is smaller in larger animals, and because energy is lost only at the surface of the body.

If the Mourning Cloak cannot reach the minimal thoracic temperature required for flight—about 18°C (65°F)—it presses the undersides of its thorax and abdomen to a warm surface and conducts heat from the substrate directly to the body. If all attempts fail to raise the thoracic temperature, the butterfly can simply return to its hibernating quarters and wait for warmer conditions. However, if the thoracic temperature is high enough, but thermal conditions are near the lower limits for controlled flight, the butterfly will fly close to the ground where warmer air temperatures and less wind resistance are encountered.

Summer offspring encounter thermal conditions different than those of spring, and they must adjust their thermoregulatory behavior accordingly. This is because high temperatures combined with intense solar radiation can produce lethal thoracic temperatures for such darkly pigmented butterflies. To avoid overheating, the Mourning Cloak shifts its thermal habitat from sunny fields to the open shade of woodlands and suburban backyards. Other heat avoidance strategies include folding the wings over the body to minimize the absorption of solar radiation, and flying early in the day and late in the afternoon—sometimes until 10:00 P.M.—thereby avoiding the hottest periods of the day.

The thermoregulatory strategies of most temperate butterflies are not as complex as those of the Mourning Cloak. Generally, temperate butterflies fly and bask at levels that correspond to their size. For example, small butterflies typically fly and bask within the protection of vegetation, which may be distributed horizontally, as in a field, or vertically, as in the case of trees. Even under hot conditions, small "field" butterflies may bask for extended periods, periodically interrupted by short, rapid flights, usually within 1 meter (3 feet) of the ground.

These flight habits place small butterflies within wind-protected layers of warm air that envelop the earth and vertical vegetations. Within such insulating "boundary" layers of air, temperatures are considerably higher and wind may be negligible. These conditions enhance heat gain and reduce heat loss. Thus, while small butterflies lose considerable body heat during flight, frequent basking within the thin boundary layer maximizes heat gain from the surrounding warmer air and vegetation, both of which may reach 52°C (125°F) or higher near the ground.

The average-size butterflies (e.g., *Colias eurytheme*) inhabit the widest range of thermal conditions, from deep tropical jungles and temperate fields to subtropical deserts and arctic tundras. They also have the greatest repertoire of thermoregulatory adaptations, including muscular thermogenesis. One unusual thermoregulatory adaptation includes seasonal *polyphenism*, found most commonly in pierid butterflies. In general, adults with heavily melanized ventral hind wings are produced when larvae are exposed to the shorter day lengths (and lower temperatures) of early spring and late fall, while adults with light yellow or white undersides are produced when larvae develop under the longer day lengths (and higher temperatures) of summer.

The melanic spring and fall phenotypes are capable of absorbing more solar radiation than their light-colored summer counterparts and can reach higher thoracic temperature faster under cooler conditions (Watt 1968). Similarly, the lighter-colored phenotypes are less likely to overheat during the summer months. Some of the *Colias* species with seasonal polyphenisms are also seasonal migrators and invade the northern temperate and Canadian zones as spring progresses. This seasonal alternation of phenotypes tracks the expected cyclical changes in ambient temperature and solar radiation, even though only one basic genotype is actually present throughout the year.

The largest butterflies have massive wingspans and slow gliding flight. The large wings of swallowtails, for example, provide tremendous lift for gliding, but they can also be a handicap on windy days. Strong, gusty winds combined with large gliderlike wings moving at a low speed can result in uncontrolled, windblown flight. The same aerodynamic problems were faced by the pilots of the Gossamer Condor, the first man-propelled glider. Smaller butterflies flying within the confines of the earth's boundary layer are not so greatly affected by gusts of wind, and many smaller species can be seen flying a few centimeters off the ground within the protection of vegetation even on days where the wind exceeds 35 kilometers ($\approx$20 miles) per hour.

Swallowtails often bask dorsally high in the trees for fifteen minutes or more under diffuse sunlight. However, these relatively massive butterflies may overheat during the essentially tropical conditions of a temperate summer. Many danaid species such as the Monarch simply do not bask under hot conditions. Instead, they protect their thorax by closing the wings, thereby reducing the thoracic temperature by up to 70 percent. In addition, some swallowtail species prevent overheating by fluttering their wings in small arcs, even while nectaring. This action creates a cooling wind over the thorax and also reduces exposure to sunlight.

Wasserthal (1975), Watt (1968, 1969), and I (1979) have each investigated various aspects of the importance of butterfly wings to thermoregulation. We found that living and freshly killed butterflies with the same physical characteristics heat and cool at the same rates and attain the same final thoracic temperatures. These results suggest that the circulation of blood through the wing veins is not a significant mode of energy exchange between wings and thorax, because dead butterflies cannot possibly circulate blood.

In fact, the bodies of dead butterflies provided with artificial paper wings "thermoregulate" as well as living butterflies! Therefore, the distribution and pattern of wing veins must have little bearing on energy transfer between wings and thorax. It appears that the thermoregulatory significance of wings is due not to physiological mechanisms but rather to the physical presence of the wings next to the body (Douglas 1979).

The typically hairy basal margins of the hind wings are modified into troughlike structures, which envelop the abdomen during basking. Scale color and pubescence in these basal areas, in combination with the troughlike hind wing margins, should also help retain body heat. My experimental results and those of Watt (1968, 1969) and Wasserthal (1975) agree with this hypothesis: higher thoracic temperatures are achieved because the properties of the basal wing areas increase energy gain and reduce energy loss.

The thoracic temperature of scaleless and hairless butterflies is reduced by 15 to 20 percent. Hairiness and minute striations and cavities within the scales are also likely to be important biophysical modifications that increase the absorptive properties of the basal wing areas. But it is the physical presence of the wings that simultaneously reduces heat loss by forming a pocket of insulating warm air between the ventral side of the wings and the surface of the basking substrate. The temperature of this insulating air pocket is increased when the butterfly is basking dorsally with the wings pressed as close as possible to the substrate.

The average butterfly is small, weighing less than a common paper clip. A small mass and large surface area allows them to gain heat quickly—and to lose heat quickly. Butterflies must, therefore, choose their flight plans carefully if they are to avoid being grounded by cold thoracic muscles in an inhospitable place. Given the high thoracic temperatures required for controlled flight, one might guess that alpine and arctic areas would be devoid of the scaly creatures. Not so, for a modest but significant butterfly fauna thrives only six degrees south of the North Pole, and at alpine heights of up to 5,500 meters (≈ 18,000 feet).

These butterflies have adapted to some ungodly thermal conditions. In the treeless, rocky desert of the High Arctic, for example, summers last four to six weeks at best, and despite continuous day, the warmest air temperature generally does not exceed 10°C (50°F). Worse still, the typically flat, rock-strewn land is buffeted regularly by strong winds, often averaging over 16 kilometers (10 miles) per hour. Because there is virtually no tall vegetation to break up the wind, its cooling effect (the wind chill effect) on small animals like butterflies is considerable.

Ecologically speaking, the alpine zone is the altitudinal analogue of the High Arctic. However, it differs from the Arctic in several fundamental ways that can affect a butterfly's ability to maintain a high thoracic temperature. First, most alpine areas receive some sunlight throughout the year whereas the latitudes of the High Arctic may not receive direct sunlight for six months. Second, alpine areas often experience upslope winds that can average 25 kilometers per hour (15 miles per hour) or more. These winds vary both daily and seasonally, and thus there is a characteristic unpredictability about them. But most importantly, the intensity of the summer sun is greater than in arctic areas regardless of the latitude of the mountain because air density and suspended particulate matter both decrease at high altitudes.

Despite these differences, High Arctic and alpine climates generally pose the same problems for butterflies: short and cool summers frequented by high velocity winds, variable periods of cloudiness, and rocky terrain with little vegetation. It is these prevailing conditions that have forced butterflies of distantly related groups to adopt strikingly similar behavioral and physical traits. Downes (1964) has commented that the thermal conditions of the Arctic are so close to the lowest acceptable limits that even small variations can have a tremendous effect on insect flight.

There are High Arctic and alpine representatives from five of the six major families of butterflies. Yet, they represent less than 1 percent of the number of species found in a corresponding temperate region. Whereas temperate butterflies vary widely in size and color, there is a striking convergence in the physical and behavioral attributes of arctic and alpine butterflies. In dorsal basking species, for example, the wing bases are commonly blackened with melanic scales. In lateral basking species, it is the ventral base of the wings that is often covered with melanic scales and hairs.

In some cases, the presence of melanic scales alters the apparent color of the butterfly. Thus, the ventral surfaces of the alpine and arctic *Colias* butterflies—the genus containing the temperate Sulphur butterflies—appear green. (See chap. 3 also). Such a combination of black and yellow scales is useful for both temperature regulation and camouflage.

In addition to wing melanization, alpine and arctic genera commonly have an extensive and uniform layer of fine hair that covers the thorax, abdomen, and the wing bases. This fine pile further increases the absorption of sunlight, and also reduces heat loss by preventing the wind from penetrating the warm, insulating layer of air trapped next to the body.

Finally, there is a dramatic convergence in the wing size of arctic and alpine butterflies when compared to the wide range of wing sizes exhibited by their temperate counterparts. Temperate butterflies range from species smaller than a child's thumbnail to the gigantic swallowtails with wing spans as big as a man's outstretched hand. By contrast, the vast majority of arctic and alpine species—even the swallowtails—have a wingspan between 3 and 6 centimeters (1 and 2 inches). Why should arctic swallowtails be no larger than a large temperate Common Sulphur butterfly? It can be argued that the reduction and subsequent convergence of wing size in all five families of arctic and alpine butterflies have two important advantages. The first is concerned directly with temperature regulation and the second with flight control.

If living butterflies are heated and cooled in a climate-controlled room, a characteristic curve is traced out on the chart paper. As the butterfly's surface absorbs light energy, the body temperature rises to some equilibrium point where the curve begins to flatten out and level off. Each butterfly has a theoretical maximum temperature it can reach under a given light source and air temperature. When the light source is removed, the body temperature drops rapidly to room temperature. Regardless of the size of the butterfly, the length of the warm-up and cool-down period, as well as the maximum body temperature reached, are functions of size, color, hairiness, and basking behavior.

With an arctic air temperature of 10°C (50°F) and a light source equivalent to a temperate sun at 10:00 A.M. on a summer day, a basking butterfly the size of a Common Sulphur (*Colias philodice*) butterfly could reach the minimal thoracic temperature required for flight in about one minute. Smaller butterflies, the size of the Spring Azure (*Celastrina argiolus pseudargiolus*) for example, also heat rapidly, but attain only 50 percent of the temperature above ambient achieved by the Common Sulphur.

This lower body temperature under the same thermal conditions reflects the Spring Azure's smaller mass, hence lesser heat capacity. Under the High Arctic and alpine conditions, the smallest butterflies would have a difficult time attaining a high enough thoracic temperature—unless they have evolved special physiological "tricks" to operate at lower temperatures. But even if the Spring Azure could fly, just a zephyr at an air temperature of 10°C (50°F) would be enough to cool them down in seconds to a point where the thoracic muscles respond sluggishly. By contrast, large butterflies such as the Monarch have a large mass and larger heat capacity, but take nearly six minutes to reach the same thoracic temperature flight temperature attained by sulphur-size species in less than two minutes.

Kingsolver (1983a, 1983b) developed and tested heat transfer models that related wing color and body hair characteristics of different alpine *Colias* populations to the pattern of body temperature and flight activity along an elevation gradient. He found that meteorological variation on a time scale of thirty to sixty seconds has important consequences for body temperature and flight activity.

In another report, Kingsolver and Watt (1983) assumed that organisms (butterflies) act as "environmental filters," which transform environmental variation into fitness, that is, reproductive success. They found that high- and mid-elevation populations of *Colias* had significantly less potential flying time than low-elevation populations, but not necessarily because of differences in cloud cover (e.g., Kingsolver 1983b). They argue that maximizing flight activity time may be constrained by meteorological variation and the need to avoid overheating. Even short-term heat shock (e.g., two hours daily at 45°C [113°F]) decreases male life span by 40 percent and female egg production by a factor of four (Kingsolver and Watt 1983).

Kingsolver (1983a) also showed that if flight time was limited, egg production may decrease in high-elevation *Colias* females. Of course, even *Colias* use heat avoidance behavior at body temperatures exceeding 42°C (108°F) (Watt 1968, 1969). Kingsolver's models of microclimate variability allowed him to accurately predict body temperature and flight activity of the population throughout the day at each site. Males tended to fly longer periods than females and spend 50 to 90 percent of potential flight time in flight when body temperatures between 30 to 40°C could be maintained.

Clearly, the rate at which the thoracic warm-up takes place is tremendously important in the fluctuating environments of alpine and High Arctic regions where the minimal thermal conditions required for flight are the rule. Faster warm-ups coupled with relatively high thoracic temperatures translate into more frequent and extended flight periods, especially during times of intermittent cloud cover that commonly prevail. A small butterfly rarely reaches the minimum thoracic temperature required for controlled flight, and when it does so, it is rapidly cooled to ambient temperature because of its small mass. Conversely, a very large butterfly takes too much precious time warming up before the next cloud blocks the sun. A medium-sized butterfly with a hairy body and black wing bases has the best of both worlds: the thorax heats rapidly to a high temperature, and it has enough mass to maintain an elevated temperature for sufficient flight periods.

Many temperate butterflies such as the sulphurs appear to be well adapted for cold environments. Why are many of these species (related to those in alpine and arctic areas) not colonizing these frigid life zones? The answer is likely due to a combination of environmental and physiological factors. Absence of an appropriate larval food plant, or competitive displacement by better-adapted species within the same group are just two possibilities that may limit access to alpine and arctic regions. However, the most important limiting factor is probably the inability of most temperate butterflies to tolerate the intense cold and dryness of the nine-month winter during the larval stage.

For example, continental arctic temperatures may average $-28°C$ $(-20°F)$ for four or more consecutive months, and the air is so dry that the insulating layer of snow common in temperate regions may be entirely absent. Furthermore, because the growing season is measured in weeks, High Arctic and alpine butterflies must diapause for two, sometimes three winters before they can metamorphose into adults. Worse, larvae are frozen completely, thawed, and refrozen unpredictably during the winter diapause, despite haemolymph that is high in natural antifreeze compounds. Thus, access to arctic and alpine environments is limited by extremely frigid winters of long duration as well as by unpredictable, cool, and brief summers.

The second reason for converging on a modest body and wing size concerns flight ability. Butterflies are not powerful insects. Their large wings and relatively small muscles create an organism which can fly only very slowly, like a glider. Gusty or strong steady winds can ground a butterfly, perhaps for energetic reasons, perhaps for aerodynamic reasons. However, smaller butterflies can fly closer to the ground, well within the thin insulating layer of warm air residing just above the tundra and rocks. They can thus avoid excessive wind resistance and maintain better flight control. Other investigators have pointed out that large wings are a distinct handicap in areas subject to frequent high winds and storms. Some groups of insects, such as arctic flies—where wings are unimportant for the passive absorption of sunlight—have wings reduced by selective forces to nonfunctional stubs, or they have evolved a wingless condition, possibly for these same reasons (Byers, personal communication).

The convergence of flight and basking behaviors in High Arctic and alpine butterflies is equally impressive. Most species fly within 50 centimeters (20 inches) of the ground. Lateral baskers in the arctic butterflies (e.g., the genus *Oeneis*) frequently will not fly unless disturbed, and then travel only a short distance, "hugging the ground" as one entomologist puts it. After it lands, it pitches over to one side, keeping the ventral surface perpendicular to the sunlight for maximum absorption of the free solar energy. This pitching behavior not only reveals the cryptically colored hind wings, but also exposes the melanized scales on the wing bases, ensuring that a high thoracic temperature is maintained for the next flight—whether it be for chasing a female, locating food, or escaping a butterfly net.

In the High Arctic, virtually all butterfly basking activity is confined to protected sunlit depressions. Within these microenvironments Canadian researchers have found that air and substrate temperatures are elevated considerably over those of adjacent and more exposed areas. Sunlit gullies become a basking haven where dozens of butterflies jostle each other for wing space on cool days. Two Canadian entomologists, Kevan and Shorthouse (1970) observed that butterflies attempt to steer themselves into these calm, wind-protected depressions—or risk being blown along uncontrollably by the stronger winds outside. They have also demonstrated that butterflies basking in these depressions on dark soil can attain thoracic temperatures 12 to 17°C (21 to 30°F) above ambient. Basking on dark substrates attenuates the cooling rates of butterflies when the sun is hidden temporarily by cloud cover. When the sun reappears, the butterflies have a thermal head start over those choosing lighter-colored substrates.

Whenever possible, High Arctic and alpine butterflies also use a behavioral temperature-regulating strategy known simply as "body contact." If substrate temperatures are higher than thoracic temperatures, the entire thorax and abdomen is pressed closely to the

warmer substrate. Heat from the substrate rapidly warms the butterfly's body. It is very likely that butterflies have external infrared,
or heat, receptors to zero in on these environmental hot spots, but
where they are located and how they function is grist for future
research. Nonetheless, body contact is a behavioral strategy most
highly developed in High Arctic and alpine species. I've seen *Boloria*
butterflies at 3,000 meters (11,000 feet) in the Rocky Mountains
literally crawling from one warm depression to another when air
temperatures were low and the sun was hidden intermittently by
clouds. Other, similar scenes have been reported from many different alpine regions.

The thermal environments of the High Arctic and alpine regions
are almost always near the physiological limit for butterfly flight
activity. Whenever the sun is obscured, even for a moment by a
passing cloud, butterflies instinctively drop to the ground as if shot
out of the air. When the sun reappears, butterflies immediately assume their basking positions and body temperatures rise quickly,
once again permitting brief flights. It is the irregularity and severity
of the High Arctic and alpine environments that act to select and
stabilize butterfly form and behavior. Thus, adult butterflies in the
High Arctic and alpine environments are under a continuous selection for size, as well as basking and flight behavior.

It is largely for these reasons that relatively small butterflies with
hairy, heavily melanized wing bases dominate the High Arctic and
alpine tundra. By carefully selecting warm basking substrates in
sunlit, wind-protected microenvironments, and by orienting the
wings to intercept the maximum amount of solar radiation, alpine
and arctic butterflies can increase and stabilize their thoracic temperature to allow frequent and extended flights.

A totally melanic wing phenotype cannot be explained fully in
terms of thermoregulatory significance. Many other selective pressures operating on the color patterns of wings must be taken into
considerations: camouflage, mimicry, and specific colors or patterns
required for mate identification are some of the most important.
Wing base melanization can be a thermoregulatory asset only if the
basking behavior allows the pigmented area to be utilized. A lateral
basker gains little thermoregulatory benefit from melanic scales deposited on the dorsal basal third of the wings. Similarly, a dorsal
basker gains no thermoregulatory advantage if the distal two-thirds

of the wings are melanized. Many factors affect the total wing phenotype. But while no single selective pressure alone can be responsible for total wing color and pattern, the basal wing areas can be modified for thermoregulatory purposes, and these areas must play an important part in the success of these small ectothermic animals (Douglas 1978).

So, the opportunity for flight activity varies widely, depending largely on the immediate thermal conditions. In the tropics, butterflies usually have a great range of potential activity from dawn to dusk throughout the year, while those butterflies inhabiting the higher altitudes and greater latitudes may be restricted by low air temperatures or sunlight intensity to only three or four hours of flight on even the warmest summer day.

Many species seem to have discrete flight times during the day in which the primary concern is either foraging for nectar, searching for mates, or ovipositing. In the Neotropics, some ithomiid butterflies (largely a forest-adapted family) forage in the open early in the morning and late in the afternoon, but seek shelter in the forest canopy when other species are just beginning their active periods. During midday, male ithomiids search for mates, while gravid females search for appropriate larval food plants, usually the Solanaceae (Drummond 1976).

Temperate species also perform specific activities at different times, although this partitioning of activity is not well understood. Perhaps nonoverlapping flight periods evolved originally to partition the food resources available, hence avoiding costly interspecific competition. It is likely that most species have preferred foraging times, mating times, oviposition times, and even specific periods for eclosion from the chrysalis. But in searching for evolutionary reasons for present-day situations, any number of reasonable hypotheses are possible, most of which will be extremely difficult or impossible to verify.

Butterflies also retire at different times of the day. Pierids typically settle well before sunset, searching for overnight sites protected by overhanging leaves or flowers. A field full of *Colias* butterflies is a jumble of flitting yellow specks during the roosting period, but a display of little yellow flags as their wings are tilted sideways to absorb the maximum amount of sunlight in the early morning hours. Those that warm the fastest to the minimum flight temperature have first choice at the nectar pots. Likewise, the first active males have first choice of the newly emerged females (Douglas 1978).

Other butterflies, particularly the Neotropical heliconiines, form permanent communal roosts to which the butterflies return faithfully each night. For example, scores of Zebra butterflies (*Heliconius charitonius*) form nightly roosts, all clinging like elegantly jeweled leaves to passionflowers (*Passiflora*, the larval host plant), or other suitable plants. As spiny larvae they sequestered toxins from passionflower vines, and hence are distasteful as adults—even poisonous to many vertebrate predators, who quickly learn that these insects are unpalatable (Benson 1971; Turner 1975).

Colony size may stay relatively stable for months as new recruits are added each week, and death occurs from disease, old age, and predation. Marked *Heliconius* butterflies have lived up to six months under field conditions (Ehrlich and Gilbert 1973). Thus, great-great-grandparents roost with great-great-granchildren—truly a family affair. Heliconiids are the Methuselahs of the butterfly world, considering that many nonmigratory temperate species have a life span of only two to three weeks at best. A roost of awakening *Heliconius* butterflies is a beautiful sight to behold: There is a progression of slowly fanning wings and a variety of basking postures as the butterflies warm up for flight in the filtered morning sunlight. Wing pumping quickly leads to slow gliding flights as the colorful butterflies drop from their nocturnal perches.

Roosting *Heliconius* seem to have limited home ranges based on the distribution of pollen sources near the roost (Ehrlich and Gilbert 1973). Roosts are typically separated by some distance, and individuals are rarely exchanged between discrete roosts. Thus, populations studied to date tend to stay remarkably stable, perhaps due to a combination of high larval mortality rates, long adult life spans, and limited pollen resources that in turn limit egg production (e.g., five to seven per day for *H. ethilla*).

Yet despite their long lives, egg-to-adult mortality in *Heliconius* is probably as high as in the checkerspot butterfly (*Euphydryas editha*) (Singer 1972). However, the life strategies of *H. ethilla* and *E. editha* are very different. Checkerspot larvae face less predation and parasitism than *Heliconius* larvae, but checkerspot adults have less dependable food resources than *Heliconius* adults. Instead, checkerspots use food stored in the larval stage to produce eggs. By contrast, tropical butterflies like *H. ethilla* can lay eggs throughout the year because adult flower resources are fairly predictable. Thus, *H. ethilla* produces eggs proportional to the amount of pollen collected daily. While *H. ethilla* ecloses with 160 oocytes, of which none is mature, *E. editha* ecloses with over 1,100 eggs of which 200 are mature (Labine 1968). Several investigators have speculated that fluctuations in population size are generally less extreme in tropical butterflies than in temperate butterflies (Brussard and Ehrlich 1970; Brussard and Sharp 1972).

Waller and Gilbert (1982) recently observed roost recruitment and resource utilization in *Heliconius charitonius*. There are two general hypotheses that explain roost behavior. One is that communal roosts protect individuals from predation because members can be quickly altered (Gadgil 1972) or because the species is unpalatable (Benson 1971; Turner 1975). The second hypothesis favored by Waller and Gilbert (1982) is that communal roosts—at least in *Heliconius* butterflies—allow information about food resources to be effectively exchanged (Gilbert 1975).

There is little data to support either hypothesis at present, but it appears that new recruits follow experienced members from roosts in order to learn the locations of pollen plants (Waller and Gilbert 1982). They have evidence that roost membership is closely tied to resource use, an important factor in long-lived butterflies (Dunlap-Pianka, Boggs, and Gilbert 1977). The limited home ranges of *Heliconius* usually but not always result in high fidelity to roost sites (Young and Carolan 1976). Roost fidelity and *trap-lining* the same nectar resources daily (Ehrlich and Gilbert 1973) helps educate predators as to *Heliconius* distastefulness.

Roosting behavior has been relatively well studied, yet its basis and purpose remain major mysteries. I wonder what would happen to a roost and resource visitation if all individuals from an isolated roost were captured and replaced by an equal number of nonrelated, laboratory-reared individuals. What would happen to the roost—would it be stable from night to night? And would butterflies learn where the floral resources are and reestablish the same trap lines?

In contrast to the heliconiines, most butterflies are highly mobile creatures, dispersing seemingly at will and without apparent direction. For example, only a few species are likely to become residents in a backyard filled with nectar-bearing plants. But why sample just a few flowers then, like Rasselas, leave paradise? Obviously, dispersal is of evolutionary importance—dispersal translates into survival for many species. Most resources are ephemeral and patchily distributed, and unless butterflies move to some extent and establish new colonies, there is always the chance that the habitat will be destroyed by a catastrophic event, or that its larval resources will be depleted by overpopulation to the point where survival of the colony is no longer possible.

I suspect that the tendency to disperse is encoded in the genes of most organisms, even in butterflies that do not exhibit much dispersal under (presently) stable environmental conditions such as those that occur in the tropics. For example, individuals of different populations of *Euphydryas editha* exhibit different dispersal rates. These differences may be the result of long-term selection pressure associated with the detailed ecology of each population (Gilbert and Singer 1973).

Dispersal is also dependent upon the phenology of the species. *Phenology* is the seasonal appearance and disappearance of a given species over the flight season. The appearance and disappearance of larval and adult resources, the annual fluctuations in environmental conditions, and the peculiarities of a butterfly's life cycle determine a species' phenology. Broods of butterflies may not always be discrete, and individuals from the end of one brood may overlap those appearing from the next brood.

Some butterflies, including many of the satyrines, have only one brood each year and are said to be *univoltine*. But many temperate species have two broods (*bivoltine*), and a few are multiple-brooded, or *multivoltine*. In the tropics, the broods of many multivoltine species overlap so that some adults are present all year round. In temperate regions, the northern populations of a species may be univoltine, the central populations bivoltine, and southern populations multivoltine or continuously brooded. Some smaller species may be multivoltine because they develop faster than larger species, but there are many exceptions to this generalization.

There are other variations on the theme of voltinism. Common temperate species like the Spring Azure have a very early spring brood that emerges from pupae that have diapaused through the winter. Some of their eggs will develop into chrysalids that will diapause and emerge the following spring, while others from the same pool of spring eggs will develop completely through to adults that are phenotypically distinct from the spring brood. This early summer brood then produces a late summer brood, which in turn produces a brood of chrysalids that overwinters. Thus, the spring brood each year consists of individuals from different generations—providing a convenient way of mixing genes within a given area annually each spring (Clench 1975).

Certain arctic and alpine butterflies (*Oeneis macounii, O. nevadensis,* and *O. jutta,* for example) take two years, possibly three, to reach the adult stage. During the winter the larvae or pupae of these butterflies are completely frozen, thawed periodically, then refrozen. High Arctic species are thus said to be *biennial:* they appear as adult with regularity only every other year. If you wish to collect these species, I advise checking ahead to make sure that adults will be flying in a given area. Otherwise, there is a chance of arriving back home empty-handed. Recently, biennialism has been documented in populations of the nymphalid butterfly *Boloria polaris stellata* (Masters 1979). Annual flights also may occur in some arctic and alpine species that are only partially biennial—a portion of the population develops within a year, while the remainder takes two years to complete metamorphosis.

Slansky (1974) established that both voltinism and the stage at which a species went into diapause were related to its choice and range of food plant (e.g., annual versus perennial, herbaceous versus woody), the number of frost-free days, and the range of frost-free days. So although there is a genetic component regulating diapause (Hayes 1979) there are overriding environmental cues that control its onset (Shapiro 1971; Hayes 1982a; Hayes 1982b). Overwintering as adults—as does the Mourning Cloak—may be the most efficient way of increasing the population early in the spring (Hayes 1982b), yet although widespread, such butterflies are only locally common.

Shapiro (1971) concludes that the West Virginia White (*Pieris virginiensis*) is univoltine largely because its exclusive food plant is the vernal, ephemeral crucifer, *Dentaria.* The genetic capacity for multivoltinism exists but its diapause as a pupae is regulated by the senescence of its food plant. Futuyama (1976) and Slansky (1976) also discovered that species using shrubs and trees as larval food plants are rarely as restricted in food plant choice as are those using herbaceous plants.

In central California, Sims and Shapiro (1983) showed that the Pipevine Swallowtail butterfly (*Battus philenor*) has a flight season extending over more than nine months (from February through November) with the major flight occurring in April derived from overwintering pupae in diapause. Most first-brood pupae undergo diapause and overwinter until the following year, but some individuals emerge to form a partial second generation that varies within and between populations. The increase in diapause, however, parallels what Sims and Shapiro describe as a seasonal decrease in young, succulent *Aristolochia* (see also Rausher 1981). Thus, the emergence of adults from the diapausing first generation is divided between summer–fall and the following spring. There is a similar discontinuity in pupal diapause period among many other lepidopterans as well.

Hayes (1982b) showed that species undergoing diapause as eggs, or reproductive diapause as adults, tend to use woody host plants. Species with pupal diapause, however, are broadly distributed over plant forms, although there is a cluster preferring woody host plants. Those diapausing as larvae are also widely distributed over plant forms; however, there is a cluster that has a herbaceous preference. Hayes (1979) contends that species with egg or pupal diapause are found largely on woody plants because eggs and pupae are best attached to woody plants. Otherwise, these stages might desiccate as herbaceous plants senesce. Finally, those nondiapausing species are confined to warm habitats and typically associated with herbaceous host plants (Hayes 1979).

Generally speaking, temperate butterflies respond to temperature and/or light cues to enter or break diapause. Decreasing temperatures and *photoperiods* (day lengths) often are the cues used by larvae, pupae, or adults to initiate diapause, while increasing photoperiods and/or temperatures cue the cells to teminate developmental arrest. Different phenologies affect interspecific potential for competition of resources, but it is likely that we will never know what the original selective pressures were that led to the evolution of diapause.

Temperate butterflies may overwinter in any stage. Generally though, diapause is restricted to one or two stages within a given species, most commonly the pupal stage (Hayes 1982b). Diapause as adults is rare among temperate species, yet the adults of several nymphalid genera hibernate in protected spots throughout the entire winter. Because hibernating butterflies may become active on warm winter days, the hibernating sites may change, resulting in dispersal. During winter diapause, the reproductive organs of the females are maintained in an immature state, probably cued by decreasing photoperiod and/or temperature. With the warmer temperatures and longer day lengths of spring, the ovarian diapause, as it is sometimes called, is terminated, and courtship takes place.

The Mourning Cloak typically chooses tight, dark places, but dozens of Milbert's Tortoise Shells (Nymphalis milberti) can be found hibernating between the loose boards of dilapidated outhouses and barns. In the fall, these overwintering adults appear to be attracted to dark openings. Sometimes, masses of hibernating butterflies can be found within ideal overwintering sites. Proctor (1976) reports that fourteen Compton Tortoise Shells (Nymphalis vau-album) approached and entered an opening only 2.0 inches long by 0.75 inch wide in a radio relay tower.

The Monarch butterfly goes into hibernation after an annual migration that brings hundreds of millions from their larval sites in temperate North America to mountainous central Mexico. Here they form massive overwintering roosts in high altitude coniferous forests within Mexico's transvolcanic belt, dominated by the oyamel fir (Abies religiosa) (Urquhart and Urquhart 1976a; Brower, Calvert, Hedrick, and Christian 1977). Tens of millions can be seen forming huge roosts on trunks and branches within the middle and lower portion of the areas of the tree crowns (see also chap. 5).

They avoid clearings and areas thinned by logging because these areas experience more extreme temperature fluctuations. At night the forest canopy is significantly warmer by 2 to 5°C (3.6 to 9°F) than surrounding clearings, especially on clear nights (Calvert, Zuchowski, and Brower 1982). This is because on clear nights the earth's surface radiates directly to the cosmic cold of outer space, and heat loss is proportional to the difference between these two temperatures. However, in the canopy, heat (radiation) from the oyamel leaves is radiated back to the butterflies, as if they were under a thermal umbrella.

As a result, the dense oyamel forests are thermally insulated. Still, freezing temperatures kill millions of hibernating Monarchs each year. Density, not canopy cover, is the key here, so forest thinning and all agricultural practices in these overwintering sites must cease immediately. Even moderate thinning can reduce forest temperatures at night by 1.91°C ($\approx$2°F), and every tenth of a degree fall in temperature near the freezing threshold kills thousands of butterflies (Calvert, Zuchowski, and Brower 1982).

Many more butterflies diapause as larvae than as adults. These include most members of the genus *Limenitis*, which hibernate as second or third instar larvae, curled up in a leaf tied together by silk strands. According to Hong and Platt (1975), diapause in *Limenitis* is "facultative" in that southern populations continue development while northern populations do not. Fritillary butterflies in the genus *Speyeria* also diapause as first or second instars within the protective leaf litter near violets, the larval food plant.

Nymphaline butterflies such as the Baltimore Checkerspot (*Euphydryas phaeton*) and other related univoltine species have an obligate diapause as fourth instars (Bowers 1978). The larvae, which emerge from huge masses of eggs numbering between one hundred and six hundred, are gregarious and live in a communal web. Before diapause they construct a small, compact prehibernation web in which they molt to the fourth instar. Prehibernation webs are always constructed above the ground and would be exposed to the drying cold of winter.

However, according to Bowers, the larvae do not hibernate in this web, but move en masse into litter below the larval food plant. Groups of these caterpillars can be found huddled together on cool cloudy days, and there is apparently limited dispersal on warmer, sunny days. After about seven days of aggregation, smaller groups (ten to one hundred individuals) split from the larger masses and move up to 3 feet away from the base of the food plant—close enough to ensure food from the perennial turtlehead plant. Here they roll up leaves and bits of debris, fastening them together with silk. Larvae may become active during thaws even in the middle of winter—not feeding of course—and can return to the dormant state without ill effect.

Diapause in the checkerspot can be "broken" by placing larvae under continuous light, and I have done the same with diapausing larvae of the Regal Fritillary (*Speyeria idalia*). However, adults are usually small—about half normal size—and many fail to expand their wings fully after eclosion because the pupal cases adhere to the expanding wings. Larvae of both species probably hibernate in the leaf litter to escape the desiccating effects of the cold, dry winter winds.

In tropical areas, some species such as Fairy Yellow (*Eurema daira*) enter *reproductive (ovarian) diapause*, a state of developmental arrest in which only the ovaries remain in an immature state. Reproductive diapause is commonest in areas with severe dry seasons. In some localities up to 50 percent of the species may be reproductively inactive. These species are responding, in a way, to the similar environmental handicap faced by temperate and arctic species—lack of water (Taylor, unpublished ms.).

By contrast, reproductive diapause is rare or absent in species that inhabit tropical or montane rain forest regions. In addition, the reproductive diapause in tropical and subtropical butterflies is categorically different from that of hibernating Mourning Cloak butterflies, in temperate regions. Old spermatophores can be found in the bursa copulatrix of tropical diapausing females, indicating that mating has already taken place—well before dry season conditions. In temperate butterflies, mating typically does not take place until the following spring. However, like temperate hibernators, the amount of fat within the *fat bodies* is considerable. And interestingly, males of tropical diapausing species may cease mate-seeking behavior during the dry season, suggesting that they too are in a state of reproductive diapause.

In deciduous tropical forests, reproductive diapause is probably necessary because of the lack of suitable oviposition sites. Diapause as adults may be more successful than as larvae or pupae because of the extremely desiccating nature of the environment, and the omnipresence of arthropod predators such as ants. Diapausing tropical butterflies must live through four or five months of dry conditions, surviving on stored fat and nectar from flowers of deciduous trees that bloom in synchrony with the dry season (Taylor 1968; Taylor, unpublished ms.).

As a rule, butterflies also diapause as larvae in alpine areas. Hayes (1980) has investigated the diapausing behavior of the Queen Alexandra's Sulphur (*Colias alexandra*), a predominantly univoltine species with bivoltine populations (Hayes 1982a). Adults are active only during a six-week period from late June to early August at elevations of 650 meters (2,000 feet) to timberline. Typical for the genus, females lay only one egg on any one plant: larvae may be dislodged or fall off the plant if they encounter another individual. Some have reported that the larvae tend to be cannibalistic under laboratory conditions.

The green larvae are difficult to see, perhaps, as Sherman and Watt (1973) have suggested, because the larvae have sacrificed some of the potential thermoregulatory advantage of dark pigmentation for a more cryptic pattern. Larvae usually rest on the midrib of a leaf after feeding, where they are well camouflaged and difficult to dislodge. Late third instar larvae become sluggish, cease feeding, crawl down and enter the litter (Ae 1958). Here they are buffered from the rigors of winter by insulating layers of leaves and snow.

Again, both photoperiod and temperature cues are likely to terminate diapause. If exposed to mean temperatures greater than 24°C during the second instar, a significant number of larvae fail to diapause (Hayes 1982a). Hayes suggests that the diapause in Queen Alexandra's Sulphur may be an adaptation with which to avoid the less nutritious and dying leaves of late summer. After resuming normal activity in the spring, the voracious larvae may consume entire, albeit small, food plants before moving on.

The range of *Colias alexandra* overlaps that of other closely related species, such as the Clouded Sulphur (*C. philodice eriphyle*), which also can utilize *C. alexandra*'s larval food plant on the high plains of the Rocky Mountains, and Mead's Sulphur (*C. meadii*), which occupies the alpine zone above timberline. In addition, the willow-feeding Scudder's Sulphur (*C. scudderi*), and the legume-feeding Orange Sulphur (*C. eurytheme*)—a migrant—also occupy the range of *C. alexandra* during the summer (Hayes 1980, 1981).

Hayes's study brings us to several important questions: What conditions limit a species to a specific habitat and geographical range, and how varied are adult movements over this range? For example, if *C. eurytheme* can utilize the leguminous food plants of *C. alexandra*, why does it not become a permanent resident of the Rocky Mountain high plains? Perhaps an answer lies in the climatic severity faced by larvae and/or adults, neither stage of which can apparently tolerate the brutality of a montane winter.

The Orange Sulphur is a generalist: it can choose from dozens of larval food plants that are perfectly acceptable to it over much of North America. By contrast, *C. alexandra* has only a few larval food plants available to it (Hayes 1982a). Furthermore, *C. alexandra*'s survival depends upon maintaining a battery of finely tuned behavioral and physiological responses that can tolerate the extreme thermal conditions—both summer and winter—as well as avoid parasites, predators, and competitors for larval and adult resources indigenous to the area.

The Orange Sulphur is also an "outsider" both behaviorally and physiologically, although this by no means implies that a population of *C. eurytheme* might not eventually adapt to an alpine habitat. Presently, under summer conditions, *eurytheme* and four other species of *Colias* butterflies can occupy similar habitats, at times foraging at the same flowers, apparently with little competitive displacement. Perhaps this is partly due to an abundant supply of nectar-bearing plants in alpine regions during the season when flight periods overlap.

Whatever the reasons for diapause, its initiation is probably not the direct result of the presence of an adverse condition, nor is the absence of diapause the result of the alleviation of those conditions (Beck 1968). Diapause has undoubtedly evolved independently in the butterfly line many times—even closely related species may diapause at different life stages and respond to different environmental cues. It is clear that by synchronizing life cycles with environmental rhythms, diapause can directly or indirectly govern much of the life history of the butterfly, from food plant choice, to reproductive strategy, dispersal, and seasonal phenology.

CHAPTER 5

Butterfly Communities: Population Structure, Dispersal, and Migration

We tend to study butterfly populations in which adults are numerous and easily accessible. However, these populations may not be representative of most natural populations (Hayes 1981). Even when hard data are available, it is often impossible to answer a plethora of questions concerning butterfly distribution, abundance, and dispersal. And of course, data can be interpreted a number of different ways so that apparent cause and effect are arranged to suit a given line of logic.

This is certainly not to belittle the efforts or intelligence of butterfly ecologists, for I count myself among their ranks. It simply means that the study of butterfly population structure—the dynamics of colonization of new habitats, the seasonal and yearly increase and decrease in populations, the reason for lack of colonization of apparently suitable habitats, or the reason for a colony's sudden extinction—often seems more of an art than a science.

Ehrlich (1979) suggests that progress in population ecology is slow because sufficient data required to answer the complicated questions involved are lacking. In some case studies the population unit (not necessarily equivalent to the species) has not been identified. That is, the size, spatial organization, and genetic properties of representative samples of plants and animals have not been properly determined. Furthermore, if we are to determine the kinds of factors that influence population numbers (and population structure in its broadest sense), then we must observe the population dynamics of a great number of butterfly species over many generations. Finally, as Hayes (1981) has pointed out, notably few life tables have been constructed for butterflies. Examples include the Cabbage White

(*Pieris rapae*) (Richards 1940; Harcourt 1966; Dempster 1967) and the Pipevine Swallowtail (*Battus philenor*) (Sims and Shapiro 1983).

Generalizing about butterfly population ecology is also difficult because butterflies occupy a great variety of habitats. These include the world's most severe deserts, the densest jungles, and arctic areas only six degrees south of the North Pole. Extreme differences in habitat can greatly affect the population ecology of even closely related species. Some species such as the ubiquitous Painted Lady butterfly (*Vanessa cardui*) occupy broad geographic ranges, whereas others are extremely localized, existing in disjunct or separated populations, such as the California Arctic butterfly (*Oeneis ivallda*), whose range is limited to alpine locations above three thousand meters in the California Sierra. The greatest diversity of butterflies however, is found within the vast rain forests of the American Neotropics.

A given species of butterfly will almost never occupy every square inch of (apparently) suitable habitat, and predicting the expected range of a species by looking at a map of "life zones" is folly at best. Ferris (1974) points out that one must look at obvious environmental factors such as annual temperature and rainfall, as well as the interaction of latitude, longitude, and altitude. Yet, because most butterflies are vagile and potentially very adaptable creatures, range extensions and new state records are reported every year.

Many complex factors of the environment must mesh with the peculiarities of a species life history before a habitat is suitable for colonization. Some species are catholic in their environmental tastes and sufficiently plastic to adapt to many different habitats. Others are limited by a specific food plant, or even certain soil types upon which certain food plants must be available for their larvae (Ehrlich 1979). Because habitats are continuously changing over centuries and millenia, the populations of many species consist of a series of distinct breeding colonies that can be envisioned as blinking on (colonization) and off (extinction) over time, like the lights on a Christmas tree.

So if we are to study the population structure of butterflies in a rigorous way, we must determine all the factors—physical and biological—that define the range of each species and regulate their numbers. Certainly climate is one of the most important factors since development must take place against the variables of rainfall, insolation, and temperature. Unseasonal cold waves or torrential rains at critical periods such as pupation or eclosion may cause the local extinction of butterfly colonies established at the extremes of their range.

Inclement weather can also weaken larvae and allow the invasion of viruses, bacteria, and fungi. These microbial pathogens may destroy whole colonies under unusual circumstances, although they usually play more of a density-dependent role in butterfly mortality. Thus, their effectiveness in reducing a given butterfly population rests on the population density—the number of individuals per unit area within the population. Other natural biological controls such as avian predators, parasitic arthropods (tiny mites and flies), and *parasitoids* (larger internal insect parasites that kill the host before eclosion) can cause significant mortality at any life stage. When a population is stressed, the combined effect of predation and parasitism can hypothetically drive a colony to extinction.

Population densities may fluctuate in a fairly regular pattern from year to year, or they can fluctuate unpredictably like the stock market. Then again, butterflies in a colony may increase to such proportions that the colony literally boils over, sending out individuals to seek suitable habitats elsewhere. If larval resources are destroyed during such outbreaks, the colony may become extinct.

Deliberate reintroduction of adult butterflies into the recovered colony's habitat may not succeed, perhaps because dispersal seems to increase in some species if adult densities are low (Brussard, Ehrlich, and Singer 1974). Released butterflies that are expected to reestablish a locally extinct colony often exhibit an *escape response* by flying toward the sun and then out of sight. However, I have found it relatively easy to establish local colonies of several species with low vagility in suitable habitats if larvae are released in sufficient numbers and protected throughout their metamorphosis by meshed screens.

In one classic continuing study (actually a number of studies in different areas initiated by Ehrlich), the population structures of different colonies of Edith's Checkerspot (*Euphydryas editha*) were found to vary according to specific habitat types.

For example, in the now famous Jasper Ridge colony of the Bay (Edith's) Checkerspot (*E. e. bayensis*), Ehrlich and his colleagues have examined the population structure since 1960. The Jasper

Ridge colony occupies an island of grassland at the top of a ridge in the Santa Cruz mountains at an altitude of about 190 meters (600 feet). This grassland area is segregated from other suitable habitats by a continuous band of chaparral and oak woodland. By marking and releasing individuals Ehrlich and his colleagues have kept track of daily and seasonal fluctuations in population, as well as approximate "home ranges" of certain individuals during the flight season and dispersal (Ehrlich 1965, 1979).

At Jasper Ridge, hibernating checkerspot larvae break diapause from late December to late January and feed on the new shoots of a native plantain (*Plantago erecta*). This plant senesces in April or May, after which the larvae diapause until new shoots appear the following winter. By early March, the larvae have pupated and the adults appear soon afterward. But Singer (1971, 1972) discovered that many larvae starve when the *Plantago* dries up and the larvae are too young to enter diapause.

The larvae that survive do so (1) because they come from eggs laid early and thus are large enough to enter diapause, (2) because they emerge from eggs laid on *Plantago* growing over gopher-tilled soil (which allows deeper root systems, greater moisture retention, and less rapid desiccation of the food plant during the hot spring), or (3) because they switched to the less common owl's clover (*Orthocarpus densiflorus*), which they consume until an appropriate stage for entering diapause is reached.

This last observation also explains why the Bay Checkerspot is found in association with serpentine soil, even though *Plantago* at Jasper Ridge also grows on sandstone-derived soil. It just so happens that owl's clover is restricted to serpentine soils (Ehrlich 1979). However, it is interesting to note that the population levels at Jasper Ridge are apparently independent of population density.

The Jasper Ridge population was originally divided into eight areas, but it was soon determined that there were three distinct and highly localized colonies—the true demographic (population) units. There is very little migration between colonies as well as emigration out of the Jasper Ridge area. Thus, each colony is apparently isolated by some unidentified "intrinsic barrier to dispersal," and few genes "migrate" as a result (Ehrlich 1961, 1979). In addition, population size in each of the three colonies fluctuates independently from year to year. Males and females were found to disperse short distances, but no behavioral devices were identified that could limit the exchange of genes—*gene flow*—within a given colony, other than male mobility.

Gilbert and Singer (1973) later studied widely disjunct populations of *Euphydryas editha* and compared individual movements and resource-use patterns of several populations to the Jasper Ridge colony. The different populations exhibit remarkably different population structures, oviposition choices, phenologies, and dispersal characteristics. The authors conclude that these differences are at least partly genetically based and the result of different long-term selection pressures on each population.

The adult population structure seems to be determined largely by the pattern of nectar resources in at least one population in Del Puerto Canyon where adult nectar resources and larval food plant resources overlap. Butterflies transplanted from Del Puerto Canyon to Jasper Ridge exhibited greater dispersal characteristics than natives, but the reasons for this are not known. The population structures of some colonies apparently are tied to "cryptic resource requirements" such as a second larval food plant—the necessity of which gravid females could not judge during oviposition.

Although individuals of *Euphydryas editha* were originally shown to be extremely sedentary (Ehrlich 1965), the mean dispersal of individuals in some populations was recently observed to change dramatically from one generation to the next. These changes in vagility appeared to be influenced by the availability of oviposition plants and adult nectar resources. Significantly greater vagility occurred in drier years when these resources were sparse (White and Levin 1981) than in wet years.

However, subsequent field observations of four *E. editha* populations in southern California indicate that the determinants of vagility are more complicated. Prior buildup of insect population numbers may cause resource depletion resulting in host plant scarcity even in years of favorable weather. Thus, the dynamic history of a population can have a significant effect on dispersal patterns of its individuals, and on the overall distribution of the organism (Murphy and White 1986).

Actually, a species population structure can only be understood within the framework of a community of plants and animals and their patterns of spatial and temporal abundance (Cody and Diamond 1975; Gilbert 1984). Evolution may occur at the species level as many contend, but it is also a community event, and species—especially species such as butterflies that interface with plants (both larval and adult food) and animals (predators, parasites, and mutualists)—cannot be studied as if they existed in an ecological vacuum. However, most studies of butterflies do just that. As Gilbert (1984) points out, understanding how a species fits into a given ecological community can help explain much about its physical and behavioral attributes.

The fundamental questions of community ecology and population biology are these: Why are there so many species, and what factors limit their distribution and abundance? As far as butterfly community ecology is concerned, Gilbert (1980) and Gilbert and Smiley (1978) have given particularly lucid reviews of the major historical and biogeographical factors that must be recognized and understood. Gilbert (1984) points out that the equilibrium theory of island biogeography was a landmark development for ecologists in general. The equilibrium theory of island biogeography proposes that on a semilogarithmic graph of species diversity versus surface area, the size of the fauna is linearly related to the size of the island.

In community ecology this theory has become widely accepted, and although it was first published by MacArthur and Wilson (1967), it was really developed by Monroe (1948) in his thesis on the distribution of Caribbean butterflies. The section concerning his musings on island biogeography were never published according to Gilbert (1984), but he did determine that equilibrium must be maintained by factors causing extinction, as well as by factors causing the formation of new species. In addition, the immigration and emigration of species must be considered. The smaller the island, and the more isolated it is from the mainland, the more likely is the hazard of extinction and the less likely the opportunity for immigration (Monroe [1948] in Gilbert 1984).

While butterfly communities—especially isolated ones such as the Jasper Ridge population of the Bay Checkerspot—may be viewed in terms of occupying "habitat islands," Gilbert argues that host specialist communities of butterflies (e.g., heliconiine and ithomiine) do not seem to follow Monroe's (1948) hypothesis. Thus, a habitat with many geographically widespread (potential) host plants does not necessarily have more species specializing on those hosts than a habitat with an equal diversity of endemic species.

Gilbert (1984) contends that broad patterns of segregation onto particular host plants are worldwide and ancient (see also Ehrlich and Raven 1965), so that local segregation onto a host plant is typically not the result of local evolutionary phenomena. However, microevolutionary interactions between species in a local area may cause the evolution of butterflies using a chemically distinct group of plants (Gilbert 1980). For example, the ithomiines feed almost exclusively on the Solanaceae. Drummond (1976) found that fifty-three species of ithomiines co-occurred on forty-four potential host plants, and the majority of these (81 percent) were restricted to a single host (Gilbert and Smiley 1978). Likewise, females of some *Euptychia* butterflies prefer small patches of host plant, yet 79 percent of fourteen *Euptychia* species from Trinidad are *polyphagous* (found on many host species) within the Gramineae. Thus, one site with one grass species supported ten species of *Euptychia* (Singer and Ehrlich [unpublished] mentioned in Gilbert and Smiley 1978).

More commonly though, butterflies utilize a more restrictive set of potential host plants than might be expected from feeding experiments (Smiley 1978a), even in the complete absence of interspecific competition (Gilbert 1984). This may, in fact, be due to biophysical or micrometeorological factors where host plants grow. However, in the controlled conditions of the laboratory, butterflies may be reared on a number of nonhost plants with great success. Shapiro (1975a) suggests that in the pierine butterflies, optimal oviposition sites are selective and limiting, but Ohsaki (1979) shows that species using the same larval host often divided up the plant on a very fine scale (Gilbert 1984).

Some populations of *Euphydryas editha* are controlled by density-dependent factors in which there is intra- or interspecific competition for larval food plants. In other colonies, parasitism and/or predation is important in population control, and little or no competition for food exists. It is clear that the taxonomic entity we classify as *E. editha* is not an ecological unit. Each population of checkerspots has its own genetic, physiological, and behavioral peculiarities, presumably under the thumb of natural selection. There are series of populations that form ecotypes roughly corresponding to the named subspecies. For example, the food plant, phenology, and adult behavior of *E. e. bayensis* (Jasper Ridge) differ from those of *E. e. luesthera* (Del Puerto Canyon). Finally, the commonly held idea that species are evolutionary units bonded together by gene flow simply does not apply to this species as a whole and perhaps does not apply to many other species as well (Ehrlich 1979).

Studies of tropical longwing butterflies in the genus *Heliconius* show remarkable stability in population structure in some but not all colonies. Benson and Emmel (1973) found similar survivorship rates and colony stability within roosts of the Daggerwing butterfly (*Marpesia berania*). The population biology of *Heliconius* butterflies and their complex relationships with their *Passiflora* host plants is particulary interesting (see also chap. 8).

Benson (1978) was the first to provide a geographical comparison of host plant partitioning in *Heliconius*. Many species of *Heliconius* require new growth (tendrils or leaf buds) to stimulate oviposition. He discovered that moderate environments were characterized by butterfly species with a higher degree of host specialization. Some species were always rare even though their larval host plant was underexploited. Drummond (1976) and Haber (1978) document similar host partitioning within communities (in Gilbert 1984).

Heliconius butterflies can be divided, roughly, into "old leaf" and "new shoot" groups (Benson, Brown, and Gilbert 1976; Benson 1978). Thus, some species lay clusters of eggs on old leaves while others lay single eggs at the tips of growing meristems or tendrils, and then only after much inspection of the food plant (Gilbert and Singer 1975; Gilbert 1979).

Food resources may also shape butterfly community patterns. Butterflies that lay more eggs may be processing more oocytes simultaneously in the ovaries rather than having their eggs mature faster (Gilbert 1984). Ovarian capacity and daily egg production in *Heliconius* are determined by the quality of larval nutrition (Dunlap-Pianka, Boggs, and Gilbert 1977; Dunlap-Pianka 1979). However, the number of oocytes per ovariole may be controlled by larval or adult nutrition (Gilbert 1984).

Gilbert has established that those heliconiines feeding on nectar alone (e.g., *Dryas julia*) are relatively short-lived (about two to three weeks) whereas those feeding on both nectar and pollen (e.g., *Heliconius charitonius*) may live for six months (Gilbert 1972; Gilbert and Singer 1975; see also chap. 8). In fact, the location of pollen plants seems to influence the daily movement patterns of communal *Heliconius* more than do larval plants (Ehrlich and Gilbert 1973).

Butterflies may coexist by partitioning time so that closely related species avoid competition for larval and/or adult food sources (Gilbert and Singer 1975; Shapiro 1975b). Also, species with novel shapes or patterns that are distasteful to predators may not have as many predators, because they have eliminated palatable butterflies that might otherwise have evolved similar shapes and patterns and hence become Batesian mimics of the unpalatable species (Charlesworth and Charlesworth 1975, and Brown and Benson 1974, in Gilbert 1984; see also chap. 7). Thus, a "model" species might also serve as a community resource for butterflies because the number of warning patterns and shapes limit the species diversity of mimicking groups (Gilbert 1979, 1984).

Insect predators (e.g., predaceous ants) and parasitoids (flies and especially wasps) can also affect butterfly community diversity and population levels (Pierce and Mead 1981). However, in at least one case, Atsatt (1981b) has established that in some species of lycaenids, attending ants must be physically present before oviposition takes place. Pierce and Mead (1981) have shown how parasitoids act as selective agents in the lycaenid butterfly–ant symbiosis. Haber (1978) suggests that rare species might be differentially eliminated by shared parasitoids, and selection might favor a shift to alternate larval host plants.

In his excellent review, Gilbert (1984) argues that population studies contribute greatly to the understanding of the community organization of butterflies and their species diversity. Of course, environmental and physiological factors also affect the population structure of a species within a community. But unless the larval and adult resources of a colony of butterflies are predictable and reasonably reliable despite climatic variations, populations tend to fluctuate, at times becoming extinct and then reestablished over time.

Euphydryas editha and *Heliconius ethilla* butterflies have been studied extensively because they are relatively sedentary species, hence easier to observe as far as population ecology is concerned. This is not to downplay the importance of these studies, which are excellent and give us a base from which new hypotheses concerning butterfly population structure and dispersal can be generated. But we need to know much more about widely dispersing species before we can make generalizations about the population dynamics of butterflies and the species diversity of local butterfly communities.

Many species of butterflies are characterized by daily dispersal behaviors that relate to mate seeking or territoriality or both. For example, Scott (1973) has documented "down-valley movement" of adult hairstreak butterflies. He defines this movement as "continuous rapid flights in a down-valley direction." These flights are apparently common in the males of western Theclini. Scott proposes that such daily flight behavior is not only correlated with mate seeking and possibly emigration, but is a feeding response. Johnson's Hairstreak (*Callophrys johnsoni*) for example, flies down-valley and feeds on mud and nectar. Scott observed that down-valley flight occurred not only during high population densities as might be expected for an emigration response, but at low population densities as well.

All species in Scott's study were "perchers," that is, males seek elevated topography or hilltops, and chase anything of reasonable size and flying characteristics that happens to come by. If it is another species or a competitive male, the intruder is driven away. If it is a responsive female, courtship may begin. In the down-valley flights, however, both males and females fly from hillsides to the valley bottom toward food resources before flying uphill again. This is qualitatively different from hill-topping behavior, a characteristic mating behavior of species whose males perch (sit and wait) for appropriate females to pass, rather than exerting themselves in an actual chase.

Recently, White (1980) documented another type of dispersal in the Anicia Checkerspot butterfly (*Euphydryas anicia*). The study took place in Colorado in the alpine zone between 3,900 and 4,100 meters (12,800 to 13,500 feet) on Mts. Cinnamon, Gothic, and Baldy. Butterflies were marked in a coded way to identify individuals and population (by peak), so that the intra- and interpeak movements could be documented. White found that population sizes varied between a low of four to six hundred on Cinnamon Mountain and a high of three to five thousand on Gothic Mountain.

In addition, and more importantly perhaps, butterflies moved up to 385 meters (1,200 feet) between captures, and transfers from one peak to another—Baldy to Cinnamon—were documented, thus establishing long-distance dispersal—2.2 kilometers (1.3 miles)—in this relatively sedentary species. This long-distance dispersal possibly took place because the Baldy population was less dense and exploited marginal habitat for the species.

Dispersal of such magnitude, of course, has the potential effect of mixing genes between largely isolated populations. Hypothetically, this should reduce the effects of *genetic drift*—a process by which rapid evolution can occur in small, isolated breeding populations. Thus, the interpeak populations of the Anicia Checkerspot would tend to be genetically homogenized if dispersal and gene exchange were frequent enough. In this case, estimated exchanges between populations were between 0.1 percent and 10 percent. This supports the contention that the successful exchange of genes by immigrants between geographically isolated colonies can produce a single population of animals with shared genes (Ehrlich and White 1980).

Shapiro (1978b) found that California populations of the Checkered White butterfly (*Pieris protodice*) differed considerably in dispersal and ability to persist as distinct colonies. Thus, the population structure of California Checkered Whites differed in this regard from the Bay Checkerspot at Jasper Ridge. Shapiro does point out, however, that some postglacial relict populations of butterflies—more typical of the transition and arctic life zones—inhabit refugia habitats on mountain tops, peat bogs, and sand areas, and this may partly explain the variation in population dynamics between the Bay Checkerspot and the Checkered White—at least in California.

Unlike California's *Euphydryas editha*, the Checkered White is a common, almost vagrant species that colonizes disturbed habitats— man-made or natural—over much of the United States. But the populations of the Checkered White in California, Kansas, and Michigan, for example, are seldom long-term, and their populations may vary greatly from one location to another, and from one year to the next. Some years they are among the most common of butterflies, yet in subsequent years they seem rare enough to be placed on the endangered species list!

Because of its long-range dispersal characteristics, Shapiro contends that the Checkered White is an unlikely candidate for speciation. Dispersal is habitual and continuous throughout the flight season, genes between "populations"—if they can indeed be called that—are effectively mixed, and reproductive rates are typically pierine, which along with some nymphalines are among the highest for all butterflies. However, we know so little about speciation in any group of animal or plants that dealing with the problem from a theoretical framework may not coincide with the reality of most speciation events. Models in general are important, though, and their use generates many useful (that is, testable) hypotheses.

In other population studies of butterflies, Watt, Han, and Tabashnik (1979) found that undisturbed montane populations of a subspecies of the Common Sulphur (*Colias philodice eriphyle*) fluctuate considerably, with the second brood numerically overshadowing the first. Although dispersal radius does not vary with brood, sex, or year, more males than females disperse in the first brood while more females than males disperse in the second.

The habitat for this montane population is marginal, perhaps in part explaining the relatively high migration rates by females in the second brood. Watt and his colleagues suggest that males need not migrate to find suitable mates in the second brood, which may explain their more sedentary behavior relative to males from the first brood, which do not have such a high density of females to choose from. As females in the second brood are more likely to be courted by males, more females tend to avoid them in a high-flying prenuptial flight that resembles a chase up a spiral staircase. This spiral chase itself may take them away from the colony—and aid female dispersal to new areas. In the first brood, males must range farther to find females, and thus they disperse more (see also chap. 7).

The range of dispersal is important, because dispersers are more likely to contribute their genes to other colonies, hence affect the course of the species' evolution. The researchers propose an "excited state" model to explain the population structure of *C. p. eriphyle*. In this model, dispersal results from a behavioral state different from that of normal flight activity. Dispersal may be affected by meteorological conditions, feeding or reproductive opportunity or the absence of such opportunity, competition, predation, or any number of other natural and physical phenomena alone or in combination. Dispersal of a colony that has reached the so-called excited state is thus a function of time, and dispersal behavior may also vary between different colonies and broods.

Even more interesting to evolutionists is the fact that the marginal populations of *C. p. eriphyle* tend to have more rare genes (*alleles*) in their gene pools. Whereas an individual may have only two different forms or alleles of a given gene (for example a gene that codes for the manufacturer of a specific enzyme involved in metabolism), sulphur

The day-flying moth *Castnia licoides* (Castniidae) has a color pattern remarkably similar to that of many nymphalid butterflies, as well as a wing venation pattern close to that of the hypothetical butterfly-moth ancestor. (Collection of S. H. Osmundson; captured at Tingo Maria, Peru, at 670 meters. Photograph by Norman Tindale.)

The ova of the Spicebush
Swallowtail (*Papilio troilus*)
are typical of the smooth,
spherical eggs of the Papilionidae. (Photograph by
Larry West.)

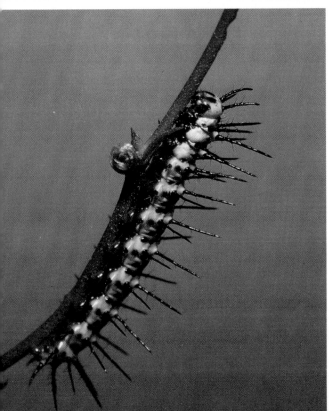

The ova of the Cabbage White butterfly (*Pieris rapae*) are characteristic of the elongate pierid egg, which typically has a series of ladderlike ridges running from top to bottom. The Danaidae, Ithomiidae, Heliconiidae, and Nymphalidae have more truncated, barrel-shaped eggs, while those of the Lycaenidae and Riodinidae are flatter and more turban-shaped, with highly sculptured surfaces. (Photograph by Larry West.)

Larvae of the Zebra butterfly (*Heliconius charitonius*) are armored with nonpoisonous spines typical of the Heliconiidae. As with nymphalid larvae, the heavy spines may serve in some defensive capacity. (Photograph by Larry West.)

The gaudy last instar larva of the Black Swallowtail (*Papilio polyxenes asterius*) resembles a small snake. As with most papilionid larvae, the surface of the cuticle is relatively smooth. (Photograph by Larry West.)

The smooth, shapely pupae of the Papilionidae are supported by a silk girdle and a cremaster imbedded in a silk pad. The pupa of the Spicebush Swallowtail (*Papilio troilus*) shown here is brown, but different shades of brown and green are known. (Photograph by Larry West.)

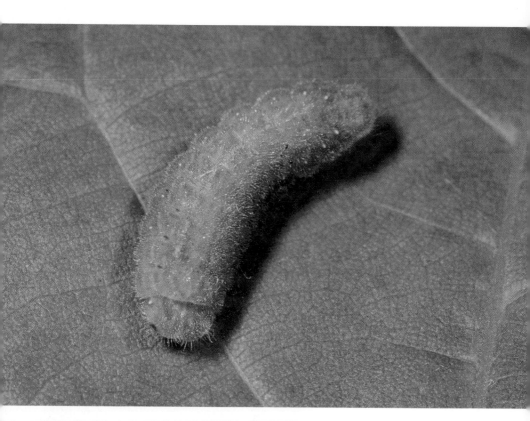

Many lycaenid larvae, such as this gerkin-shaped larva of the Banded Hairstreak (*Satyrium calanus*), are covered with minute hairs that among other functions, may help the larvae to blend in with the leaf surface. (Photograph by Larry West.)

The Baltimore Checkerspot (*Euphydryas phaeton*) emerges within ten days after forming its harlequin-colored pupa. As with other nymphalids, the pupa hangs from a ventral cremaster whose hooks are sunk into a pad of silk bound to the supporting substrate. (Photograph by Larry West.)

The danaid pupa of the Monarch butterfly (*Danaus plexippus*) is smooth and elegant with its gold-leaf trim. (Photograph by Larry West.)

The transparent cuticle of the Monarch pupa shows well-developed wings only a few days prior to eclosion. (Photograph by Larry West.)

The body parts and wings of the Monarch are very distinct only a few hours before the adult emerges. The pupal skin is brittle and easily cracked. (Photograph by Larry West.)

Within minutes after eclosion, the soft-winged adult Monarch has secured a perch from which it can safely expand its wings with haemolymph. Note that the empty pupal skin has no trace of the colors evident in the newly formed pupa. (Photograph by Larry West.)

Although the wings of this female Monarch are fully expanded within minutes or so, they remain soft and cannot support the insect in flight for at least an hour. (Photograph by Larry West.)

The dorsal basking orientation is depicted here by a male Julia butterfly (*Dryas julia*), a heliconiid with a relatively short life span. Notice the overlapping wings and the raised abdomen. (Photograph by Larry West.)

Body basking is a common thermoregulatory behavior of lycaenids. Here, an American Copper butterfly (*Lycaena phlaeas*) basks as it nectars on a composite flower. (Photograph by Paul Douglas.)

Lateral basking is also a common thermoregulatory strategy among the Lycaenidae, especially the hairstreaks and elfins. This Banded Hairstreak (*Satyrium calanus*) imbibes nectar and palpates the flower with its antennae as it basks. (Photograph by Paul Douglas.)

Several North American nymphalids such as the Mourning Cloak (*Nymphalis antiopa*) hibernate as adults. This California specimen emerged from its hibernating quarters in a splintered log, and assumed a lateral basking position prior to shivering. Note the thick pile of hair and scales that insulates the wing bases and thorax, and allows the absorption of more sunlight. (Photograph by Matthew Douglas.)

Summer form of the Question Mark butterfly (*Polygonia interrogationis*) extracting nutrients from fox feces —an excellent illustration of "alternate food sources" often used by nymphalids. (Photograph by Larry West.)

A crowd of Coral Hairstreak butterflies (*Harkenclenus titus*) forage at butterfly weed (Asclepiadaceae). So engrossed are these lycaenids in their foraging activities, that they can be gently coaxed onto a finger or captured neatly with a pair of forceps. (Photograph by Paul Douglas.)

Papilionid butterflies such as this mob of Tiger Swallowtails (*Papilio glaucus*) are common visitors at mud puddles and urine patches, where they probe for high concentrations of sodium. (Photograph by Paul Douglas.)

This last instar larva of the Red Admiral (*Vanessa atalanta*) is parasitized by minute wasp (Hymenoptera) larvae. Several parasitic larvae have emerged from the rear of the caterpillar, and will pupate on its skin. (Photograph by Matthew Douglas.)

Many nymphalid larvae, such as those of Milbert's Tortoise Shell (*Nymphalis milberti*) are gregarious or communal. Here a number of larvae attack stinging nettle, a common food plant. (Photograph by Larry West.)

The larvae of the Giant Swallowtail (*Papilio cresphontes*), like all true swallowtail larvae, evert foul-smelling osmeteria when threatened. Larvae of the Giant Swallowtail are known as "orange dogs" by citrus growers in the south, where the larvae occasionally reach pest levels. (Photograph by Larry West.)

Yellow males and melanic females (dusted with yellow scales) of the Tiger Swallowtail (*Papilio glaucus*) are fairly common in the northern transition zone of southern Michigan. North of this zone, the yellow female predominates, and south of this zone, the melanic female type increases in frequency. Presumably the melanic forms are Batesian mimics of the distasteful Pipevine Swallowtail (*Battus philenor*), which is a common southern species, but uncommon to rare in transitional zones such as southeastern Michigan. (Photograph by Matthew Douglas.)

Unless disturbed, these mating Wood Satyrs (*Euptychia cymela*) will rarely fly. Note the apparent absence of the forelegs, which are reduced to short stubs tucked tightly against the body. (Photograph by Paul Douglas.)

Like many tailed lycaenids, the Striped Hairstreak (*Satyrium liparops*) rubs its hind wings back and forth while perching, giving the impression of a "false head" to would-be predators. Males use such perching sites to search for passing females, as well as for basking. (Photograph by Paul Douglas.)

A mating pair of the Great Spangled Fritillary (*Speyeria cybele*). Note the color difference between the darker female (on the left) and the lighter male (on the right). (Photograph by Paul Douglas.)

These pupae of the Cabbage White butterfly (*Pieris rapae*) show the range of color polymorphism common in the Pieridae and Papilionidae. Pupal polymorphism can greatly affect the degree of camouflage on pupal substrates of different color. (Photograph by Robin Franklin Bernath.)

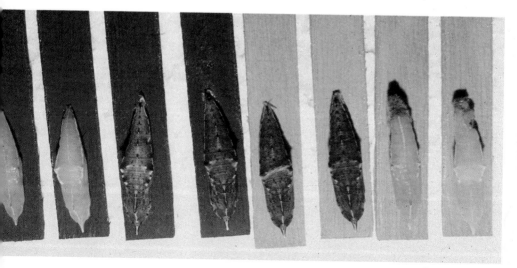

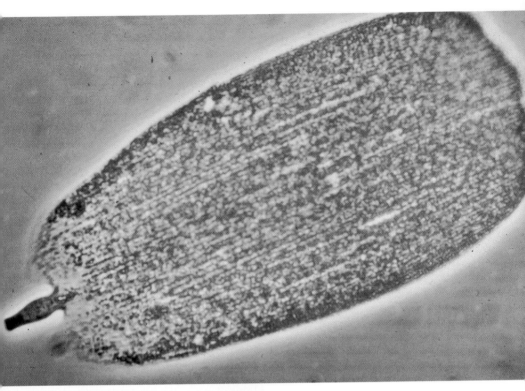

Under ultraviolet light an ultraviolet-reflecting scale of the Orange Sulphur (*Colias eurytheme*) appears as brilliant as gold. (Photograph by Robert Silberglied/Orley Taylor.)

This flashing male Orange Sulphur (*Colias eurytheme*) is trying hard to entice a stationary female, but the elevated abdomen and fluttering wings show she is not interested in any procreative activities. (Photograph by Robert Silberglied/Orley Taylor.)

populations in marginal areas may have a total of six alleles. Perhaps such genetic diversity is important in marginal populations; the extra alleles might be carried like so much excess baggage "just in case," as a parent might say.

Many butterflies disperse throughout the years in a seemingly random manner. That is, their range of distribution varies from year to year. Still other butterflies undertake incredible migrations, covering hundreds, even thousands of miles en masse or as individuals. *Migration* differs from dispersal in that dispersal is often undirected, whereas migrations usually take place in directions that are seasonally determined and predictable. Migration, like hibernation, is probably an evolutionary survival response to inhospitable conditions. For example, some species migrate to northern areas during the spring and summer, and their progeny return to southern areas to overwinter as adults.

If a species cannot tolerate extreme dry seasons in the tropics or severe winters in temperate zones in any developmental stage, then one effective alternative—indeed the only one—is to seek greener pastures elsewhere. Species that are good long-distance migrators, such as the Painted Lady butterfly (*Vanessa cardui*) and the Monarch butterfly (*Danaus plexippus*) may successfully colonize vast areas of the temperate zone during the summer. A fall brood then migrates south. These migrating species are some of the most abundant, widely distributed, and phenotypically least variable of all butterfly species.

The Painted Lady migrates in tremendous numbers, en masse, while those such as the Dainty Sulphur (*Nathalis iole*) migrate as individuals. Some species migrate in one direction and their offspring return in the opposite direction, while others migrate in only one direction during "outbreaks" and neither they nor their progeny return.

No one knows the precise reason for migration in the two-hundred-odd butterflies known to migrate. Most migrators are tropical or obviously of tropical origins (Emmel 1975). Some migrate in response to photoperiod and temperature cues that signal the onset of fall and winter. Others migrate possibly because adult food sources are scarce or lacking, or because of overcrowding during the larval stage. Whatever the reason, the flight behavior and often the physiology of the migrating individuals change remarkably, even though nonmigrants and migrant forms can be obtained from the eggs of a single female if the environmental cues received by the developing larvae mimic the cues received in the environment during the summer and the fall.

Outside of a few well-known migrators such as the Monarch, it is almost impossible to predict a butterfly migration. For this reason hard data concerning migration direction, condition of individuals, and numbers of individuals are lacking. Thus the record of butterfly migration is replete with anecdotal footnotes and short papers that relate an observation but do not advance a hypothesis.

For example, the Gulf Fritillary butterfly (*Agraulis vanillae*), is sometimes reported as a "breeding migrant"—that is, one that migrates far out of its range. It establishes colonies up to 1,000 kilometers (600 miles) north of its permanent habitat. During the winter, breeding colonies are restricted to the subtropical margins of the United States and southward. But the Gulf Fritillary is a tough, hard-flying butterfly, and it is commonly reported in Missouri and Kansas during the summers where its larvae may denude local passionflower vines (*Passiflora*) (Howe 1965).

Is the Gulf Fritillary a true migrant? Yes, because like the Monarch, some individuals stray northward—riding the southerly winds of spring—and reproduce, even though their offspring cannot survive a killing frost. Worn, migrating Gulf Fritillary butterflies are never seen early in the spring, and in some years the migrations are small perhaps because of adverse environmental conditions in the southern breeding areas. In these years, few if any migrants may reach the northern states.

The migrations of the Great Southern White butterfly (*Ascia monuste*) are even more of a mystery than those of the Gulf Fritillary. Whereas the Gulf Fritillary migrates as individuals, the adults of the Great Southern White often fly in a tight cylindrical form that seems to undulate up and down as it passes overhead. Such a snake-like formation may contain millions of individuals. During the final hours of the migration the flight formation breaks down and the mature females locate larval food plants and oviposit.

The Neotropical pierid butterfly *Kricogonia castalia* also takes part in rare migrations. This species is common in northern South America, Central America, Mexico, and occasionally in the subtropical region of Texas and Florida. Byers (1971) reported a tremendous migratory flight of these butterflies in Central Tamaulipas,

Mexico, on July 11, 1961. The migration was moving from north-west to southeast toward the Gulf of Mexico, and the width of this immense stream of butterflies measured about 90 kilometers (53.5 miles) with an undetermined length. The butterflies, like many migrants, flew within 2 meters (6 feet) of the ground, and usually within a meter (3 feet) of the surface.

For each mile Byers estimates eleven hundred butterflies crossing an average mile per minute (1.7 kilometers/minute). Another vast migration of these Neotropical pierids, flying east to west, was reported to obliterate the highway between Ciudad Mante and Ciudad Victoria, Mexico, on October 23, 1963 (Howe 1964). During both reported migrations the weather was hot, partly cloudy to sunny, with little wind.

Migrating butterflies may be channeled in a certain direction by impassable geographical features such as forests, mountains, or deserts (Douglas and Grula 1978). Migrations may also be affected by the way the land is oriented in relation to vast bodies of water such as the Great Lakes or the oceans. For example, Urquhart and Urquhart (1976b, 1976c) followed annual migrations of the Monarch, the Gulf Fritillary, and the Cloudless Sulphur (*Phoebis sennae eubule*) along the coast of northern Florida. The migrants passed through in the greatest numbers between October 20 and 25, following the coastline in a westerly direction. The Cloudless Sulphurs would fly northward only a few centimeters above the water, then abruptly alter their course and pursue a westerly direction.

The Urquharts suggest that the records of flight direction given for the Gulf Fritillary by Williams (1930) are confusing because flight orientations can be recorded in every direction depending on the geographic location of the observer. Overall, these butterflies appear to move in a southerly direction, but a westward component may be introduced at coastal areas where the coastline proceeds to the west (Urquhart and Urquhart 1976b).

There are, of course, many other anecdotal stories of butterfly migration, but at least three species of butterflies with very different patterns of migration are worth considering in greater detail. These are the Painted Lady, the Monarch, and the Dainty Sulphur. Without annual migrations the geographic ranges of these species would be severely limited. Yet each has vastly extended its range into temperate regions if for only a few months every year. None, however, can survive northern winters. So we must ask ourselves, "Of what value are migrations if migrants or their offspring cannot survive to reproduce—why not remain in a more hospitable climate where reproduction can continue the year round without interruption?"

The Painted Lady is one of the most cosmopolitan and periodically abundant of all butterflies. It inhabits all but two continents—Antarctica and South America—at some time during the year, preferring open, sunny areas throughout its distribution but not tropical forests. Despite its incredible geographic range, possibly only a few nonmigratory colonies exist, and no subspecies is officially recognized, although a closely related species (*Vanessa kershawi*) occupies Australia, New Zealand, and some of the Pacific Islands (Williams 1970).

The cosmopolitan habitat of the Painted Lady and its apparent lack of distinct, geographically defined subspecies suggests that gene flow and exchange is large, approaching at least a continental scale. Over most of its temperate range it is a frequent but temporary summer visitor, and its numbers fluctuate dramatically from year to year. Winter survival has not been documented where killing frosts are experienced.

In the spring, Painted Lady adults from the Mediterranean-like climates move singly or in huge migrations northward. Several large migrations have been recorded in North America. In June, 1952, for example, a migration lasting for seven days was observed in the town of The Pas, Manitoba. These butterflies were in relatively new condition, contrasting sharply with the worn, tattered butterflies of a much more massive migration through the same area ten years later. Neither of these northerly spring migrations produced a large southerly flight during the fall as was expected (Krivda 1976).

Another massive migration of older Painted Lady butterflies was reported by Brown (1974) a few miles southeast of Colorado Springs, Colorado, on April 28, 1973. Brown estimated 100 to 150 individuals passing directly in front of him for each mile for a distance of 120 kilometers (70 miles). Concentrations on lawns were estimated to exceed 500 per acre ($\approx$2.5 hectare); most were feeding on dandelions and apple blossoms. The migration continued for nearly thirteen days, after which time another huge migration of newer-appearing butterflies passed through, covering lawns with nearly 1,000 butterflies per acre—a concentration that persisted for nine days. Clearly, the numbers of butterflies in some migrations must be in the hundreds of millions.

Painted Lady migrations occur with equal or greater magnitude from North Africa all the way through the Arctic Circle in Scandinavia and Finland. Furthermore, there are many records of Painted Ladies flying thousands of miles out to sea, between, for example, the Mediterranean and the West Indies. The cosmopolitan distribution of the Painted Lady reflects the catholic tastes of its larvae— over a hundred different food plants, primarily in the daisy (Compositae), mallow (Malvaceae), and pea (Leguminosae) families, have been recorded. Larvae reach pest proportions on several domestic American crops such as soy beans, maize, beans, and sunflowers.

During the summer in North America, the Painted Lady may be captured at latitudes over sixty degrees north. The parent swarm apparently originates in Mexico where it is assisted in its migration by the prevailing south winds of spring. The flight direction is typically north or northwest, with a few traveling northeast. Butterflies usually fly below 3 meters (10 feet), and swarms estimated to contain up to 300 million individuals have been documented. Yet, all progeny of these northward migrating individuals must perish with the onset of winter, unless there is a return, southward migration.

There is some evidence of southerly migration in the fall. For example, Emmel and Wobus (1966) noted a southward migration during the late summer and fall of 1965 near Florissant, Colorado (by coincidence also a famous butterfly fossil site). The migration took place intermittently on August 22 and 28, and again between September 1 and 19. The migration rate on August 22 was six hundred butterflies crossing a 6-meter (20-foot) line per hour, and this continued from dawn to dusk. In other records, butterflies have been observed in equally astronomical numbers, flying south despite a westerly wind. Although wind does not always influence the direction of migration, the advent of a cold wave is known to "dampen" the migratory activity, possibly for thermoregulatory reasons.

As with most migratory species, no one is certain how specific flight orientations are achieved and maintained. Prevailing winds, polarized light, and magnetic fields and magnetite crystals in the butterfly's brain have all been implicated. Perhaps orientation requires a combination of environmental stimuli. For example, Painted Lady butterflies are quite muscular for their size, and I have seen them struggle upwind in a flight pattern as close to the ground as possible. Estimates of ground speed range from 20 kilometers per hour (12 miles per hour) against a strong headwind to 33 kilometers per hour (20 miles per hour) with the wind.

In some species, different behavioral and physiological phenotypes are likely present throughout the year: a spring form that migrates northward, a more sedentary summer form that may continue the northward dispersal, and at least in some years a fall form that migrates southward. This seasonal polyphenism at the physiological and behavioral level can be explained by genes that are directly or indirectly responsive to specific environmental cues. But how this is done—if indeed this is how it is done—remains a mystery.

Perhaps the most famous migrant species, and easily the best known of American butterflies, is the Monarch. The larvae feed on the milkweeds (Asclepiadaceae), which include about 108 species distributed over much of North America. Because their breeding range spans several life zones, one or more broods may be produced each year. The rate of larval development, of course, is dependent on thermal conditions.

Over the past millions of years the tropical ancestors of the modern Monarch evolved a migratory response (as well as a diapause and hibernation) that now allows their descendants to escape the cold and dryness of a temperate winter. However, the summer form is quite different—both physiologically and behaviorally—from the migrating fall form. For example, butterflies of summer broods fly singly, disperse randomly (not in a specific direction), fuel themselves with carbohydrates, and are reproductively active. They also appear to lack the muscular thermogenesis capacity of fall forms (Kammer 1971).

By contrast, the fall migrants are typically in a state of reproductive diapause, usually do not search for mates, have the ability to generate heat internally by shivering if body temperature is too low to initiate coordinated flight, and have huge fat stores that they use during the six-month-long overwintering period. Finally, fall migrants somehow orient in a south to southwest direction, fly thousands of miles in a matter of weeks or months, and then locate and overwinter in the same general roosting sites by the millions without having ever seen these places before!

During the fall migrations, butterflies may shiver, elevating thoracic temperature up to 8°C (14°F) above ambient (Kammer 1970). Both basking and shivering help them maintain coordinated flight under cool conditions. But orientation has been a much more difficult problem to tackle. Baker (1968) postulated the evolution of sun orientation for dispersal and migrations, and such orientation is apparently used by some pierid and nymphalid butterflies. Since the Monarch is also diurnally active, it might also use sun compassing. In fact, most diurnally active butterflies will move in accordance with the sun's position, often as an escape response.

Kanz (1977) suggests that using the sun is an appropriate escape avenue because it is the most prominent cue available, it is fast because it offers a linear escape route, and finally, predators should have a more difficult time following a butterfly obscured by the direct rays of the sun. Kanz found that fall migrants oriented to the sun's azimuth when the sun was visible regardless of the experimental conditions. And since most Monarchs migrate from 10 A.M. to 2 P.M.—the hottest temperature of the day during the late summer and fall—the overall direction is south-southwest. This is because the sun is tracked in an arc of only sixty to seventy degrees rather than one hundred eighty degrees if flights continued from dawn to dusk. Thus, restricted sun orientation could allow individuals to be funneled predominantly in one direction.

Monarchs may also be steered southward by advancing cold fronts that may cause them to settle in small clusters at specific "Monarch pit stops" along the migratory pathway. These pit stops may be used regularly year after year, such as the one at Point Pelee, Ontario, or only once or twice during a ten-year span. And recently, too, there is evidence of minute magnetite crystals in Monarch brains, which suggests that they are using magnetic lines on the earth for orientation as well. However, magnetic field lines are locally much more irregular and therefore likely to be a less error-free source for direction than would be the sun. Nonetheless, magnetic orientation would explain how Monarchs can navigate on heavily overcast, even rainy, days. It is also possible that Monarchs navigate using thermal sensors between the antennae on the vertex of the head—perhaps on the chaetosemata. It appears then that fall migrants use a number of environmental cues, singly or in concert, depending on the environmental conditions.

Immense numbers of Monarchs can be observed migrating down

well-defined flyways. One of these is along the eastern shore of Lake Michigan, where hundreds of thousands perish annually on the cold autumn sands washed by stormy waves. Temporary local congregations can be found across much of eastern North America, save for the highest mountain ranges. Monarchs can soar over tall buildings effortlessly, gaining elevation as rising air from thermal lifts pushes against their beautifully constructed gliding planes. At the University of Kansas, I have seen Monarchs glide over the football stadium, then zoom in spiral formation over Snow Hall, using heated columns of air channeled between projections of the building as thermal lifts.

By late fall, millions of Monarchs are approaching the overwintering sites in Florida, Texas, California, but especially at protected sites within the Central Transvolcanic Range of Mexico. Most but not all Monarchs overwinter in reproductive diapause. Brower (1961) discovered that breeding populations of Monarchs exist in Florida during midwinter, spring, and summer months. Other breeding colonies have been reported in southwestern Arizona (Funk 1968) during December and March of 1965. The fall migrants that do not breed generally travel from the northeast to the southwest, which explains in part why migrants are not as common in peninsula Florida (Urquhart and Urquhart 1976c). However, tremendous numbers move west along the Gulf coast of northern Florida. It is likely that some migrants pass through peninsula Florida, across the Keys and western Cuba, to overwinter in Guatemala and the Mexican Yucatan.

The vast majority of migrants, however, congregate in huge overwintering roosts in California and Mexico. The coastal sanctuaries within the United States are located from Los Angeles to north of San Francisco, where the butterflies form congregations in protected groves of Monterey pines and eucalyptus trees that remain moist and above freezing throughout the winter. These trees are probably chosen because the leaves are narrower, easier to grasp, and thus can support many more butterflies than broad-leaf deciduous trees. Some conjecture that roosting trees produce volatile plant compounds that can be detected by the migrants. At Pacific Grove, California, an entire tourist industry has developed around the migrants, which includes an annual festival and a five-hundred-dollar fine for anyone caught molesting Monarchs.

In late 1974, Bruggar discovered a vast overwintering colony located on the slope of a volcanic mountain in the northern part of the state of Michoacan, Mexico. The 8-hectare (20-acre) roost was established predominantly in the coniferous oyamel fir tree averaging approximately 24 meters (80 feet) in height and growing at 2,800 meters (about 9,200 feet) (Urquhart and Urquhart 1976a).

The discovery of this overwintering site was the culmination of years of tagging, releasing, and recapturing migrant Monarchs by the Urquharts. Monarchs are packed together in such numbers that the underlying bark is not visible. The top portions of the trees are usually free of butterflies as are the lower portions of the trunk, probably because thermal conditions are much more extreme here due to wind exposure and freezing temperatures respectively (Calvert, Zuchowski, and Brower 1982). Because of their location within the canopy, the body temperatures of roosting butterflies are usually above freezing, and on sunny days the butterflies bask at the site and may even fly to lower elevations, foraging for nectar or drinking water. By mid-March, about 75 percent of the butterflies have left the colony for the northward return flight.

On December 31, 1976, Brower and his colleagues independently located a massive overwintering colony of Monarchs estimated to contain thirty to one hundred million butterflies, also in the central mountains of Michoacan, and possibly the same overwintering site first seen by Bruggar and described by the Urquharts (Brower, Calvert, Hedrick, and Christian 1977). Brower's site, designated Site Alpha, is actually only one of several overwintering colony sites now known in the Mexican mountains. Others undoubtedly exist in areas with suitable overwintering environments.

Site Alpha is 165 kilometers (100 miles) from the sea, comprises approximately 1.6 hectares (4 acres), and lies at elevations between 3,000 and 2,900 meters (9,900 and 9,500 feet). Most of the estimated twenty-four hundred trees are firs, (*Abies religiosa*), cypress (*Cupressus lindleyi*), pine (*Pinus ayacahuite*), and two small broadleaf trees. Brower reports that the distribution of the Monarchs in the coniferous trees appears adapted to avoiding wind, frost, and snow. Minimum temperatures near the roost ranged from 5.5 to 8.8°C (42 to 48°F) and maximum from 13.3 to 15.5°C (56 to 60°F), a thermally stable range considering the diverse weather conditions experienced within the Central Transvolcanic Range.

In effect, the butterflies—in reproductive diapause—are hibernating like their temperate nymphalid counterparts, but at higher ambient temperatures. It is cool enough to reduce metabolic rates considerably, moist enough to prevent desiccation, yet protected enough to avoid freezing air and heavy snows. The forest modifies the entire thermal climate and in turn, the presence of millions of butterflies modifies the forest both thermally and physically. Some branches are so heavily weighted with dormant butterflies that they break, sending thousands of butterflies crashing to the ground. At temperatures below 1.5°C (35°F), butterflies dislodged from the roosts flap helplessly on the ground, unable to fly. If shivering fails to warm the flight muscles sufficiently to permit flight, the butterflies often perish from freezing temperatures, dew, or snow at the base of the trees. Dislodged butterflies usually attempt to climb up off the ground (Calvert, Zuchowski, and Brower 1982).

Most roosting females are virgins and remain reproductively inactive until photoperiods begin lengthening and warmer temperatures arrive by late January and early February. By mid-February, swarms of mating Monarchs fill the air, and by late March the entire overwintering brood disperses. After mating, egg formation takes place as yolk protein (vitellogenin) is synthesized by the fat body cells under direction of juvenile hormone (Pan and Wyatt 1971). Egg formation does not take place until day lengths reach eleven to twelve hours and temperatures approach 20°C (68°F) (Barker and Herman 1976).

As with the mixing of migratory Painted Ladies, the overwintering Monarchs provide even greater opportunity for gene exchange and homogenization of the species phenotype. Obviously, the seasonal polyphenic differences between summer forms and fall migrating forms are genetically based, and the annual formation of huge overwintering colonies must serve to keep the genes of the migrating complex intact.

As the colony breaks up, individual Monarchs fly off in a northern or northwestern direction, and lone females oviposit on appropriate milkweed species as they travel northward. It is not known how far these returning migrants reach—perhaps in some cases all the way back to their points of origin. However, it is more likely that the progeny of migrants continue the spring migration northward in leapfrog fashion. In addition, spring migrants may be joined by individuals that remained and bred in southern Texas and southern Florida. Monarchs in these areas are uncommon during June and July because most of the migrants have dispersed to the north (Neck 1976b). It is likely that returning Monarchs ride the persistent and strong southerly winds northward during the spring. Perhaps they also use magnetic compassing as well as a negative sun orientation since the sun is now at their backs.

What has been learned about the life history of the Monarch butterfly could fill an entire bookshelf. However, many mysteries remain. How do the butterflies orient south in the fall, then north in the spring? How long does the fall/spring migratory form live—nine months, or longer as speculated? Do fall migrants ever return to their original birth sites in the spring? How did such an incredible migratory response evolve, and finally, how is the physiological and behavioral polyphenism controlled by the genes? These and many more questions will take batteries of scientists years to answer, and in the process many more questions will be generated.

There are other, lesser-known butterflies with migratory phases that give us an idea of how organisms might permanently expand their geographic range, perhaps splitting into two or more distinct species from one parental species in the process. A true *range* or *ecological expansion* involves adaptations to cope with barriers that formerly prevented occupation of a given habitat (Lewontin and Birch 1966). Most range expansions are geographical range expansions that result from the removal of physical barriers that formerly prevented occupation, or to the sudden alteration of the environment. Geographic range expansions are relatively common phenomena, particularly with weedy or pest species, whereas ecological range expansion and documentation of the evolution of adaptations that allow the expansion to take place are rare.

For butterflies, seasonal activity is regulated to a large degree by photoperiod and temperature, both of which are often the environmental cues that trigger hibernation, migration, diapause, and the seasonal polyphenism for many behavioral and physiological characteristics. Photoperiod in particular is a relatively reliable and noise-free indicator of cyclical changes in the thermal environment, at least in nonalpine areas.

One adaptive polyphenism common especially among sulphur butterflies involves seasonal changes in adult wing pigmentation.

The Orange Sulphur (*Colias eurytheme*) shows a photoperiodically regulated deposition of melanin on the ventral surface of the hind wings (Ae 1957; Hoffman 1973). Spring and fall forms have greater numbers of melanic scales, particularly in the basal areas next to the body, whereas summer broods have few melanic scales. Of course, fall and spring forms can reach thoracic temperatures required for flight under low solar angles and cooler temperatures faster than nonmelanic forms (Watt 1969).

The geographically distinct midwestern population of the Dainty Sulphur (*Nathalis iole*), a subtropical coliadine, has recently evolved a similar sensitivity to photoperiod (Douglas and Grula 1978). Larvae exposed to short photoperiods (ten hours of light) develop three times the number of basal melanic scales as those from the same female exposed to long (sixteen-hour) day lengths. Melanic adults heat significantly faster and to higher equilibrium thoracic temperatures than their immaculate counterparts, giving melanic forms a thermoregulatory advantage in seasonally or geographically cooler environments (spring, fall, northern areas) and immaculate forms the advantage in hotter environments (summer and subtropical areas).

Only short photoperiods—not reductions in temperature—are required to produce the melanic phenotypes in these tiny sulphur butterflies. Like most if not all Coliadinae, the Dainty Sulphur is a lateral basker and never uses the dorsal basking position. Thus only melanic scales on the ventral surface are of thermoregulatory significance. Overheating can and does occur in midwestern regions during the summer as temperatures near the surface exceed 50°C (122°F), making the immaculate form the most fit for thermoregulatory reasons. Conversely, melanic forms are capable of more extended flights during spring and especially fall, at which time the Dainty Sulphur is one of the last butterflies in flight.

Now for the rest of the story. According to Clench (1976b), North American populations of the Dainty Sulphur can be divided into three distinct regional segregates (southeastern, midwestern, and southwestern) between which little or no gene flow occurs. The midwestern segregate ranges from Guatemala and Mexico to southern Ontario and Manitoba. It undergoes an extensive seasonal migration that originates in southern Texas in early spring and typically terminates in southern Canada by late summer. The migration is actually composed of successive broods of butterflies extending their northern range as long as life span and the thermal environment permit. Each generation probably moves at least several hundred miles, aided by strong southerly winds, and channeled across suitable prairie habitat between the Rocky Mountains and the less penetrable eastern deciduous forests.

The midwestern segregate also is well known for its seasonal altitudinal migrations in the Rocky Mountains to nearly 3 kilometers ($\approx$10,000 feet). These individuals, like their prairie counterparts, are cold-tolerant as adults and have been captured under freezing conditions in December at Colorado Springs, Colorado. By contrast, the southeastern segregate centered in peninsula Florida and the southwestern segregate centered in southern and Baja California do not migrate and are apparently cold-intolerant, although they also exhibit the seasonal polyphenism. Finally, there is an isolated, near-equatorial population of the Dainty Sulphur (or a close relative) that exhibits only the melanic phenotype (Shapiro, personal communication). This relict population is presumably multivoltine and probably does not diapause.

The Dainty Sulphur migration appears to be a general adaptive dispersion in all directions that is channeled throughout the Plains because of physical barriers—mountains on the west and forests on the east. Undoubtedly, strong spring winds assist the migrating individuals as they fly within 60 centimeters ($\approx$2 feet) of the ground. Weins (1976) points out that dispersal—including long-distance migration—should be well developed in species that occupy patchy habitats. The dispersal characteristics of the midwestern Dainty Sulphur with its choice of patchily distributed, ephemeral composites fits this hypothesis.

So apparently we have a single species, with three reasonably distinct segregates, at least one of which has already acquired some of the behavioral and physiological characteristics needed to complete a permanent ecological range expansion. These characteristics include the well-developed thermoregulatory and dispersal abilities of adults. And although the Dainty Sulphur cannot diapause (as far as is known) anywhere in its range, it nonetheless has the ability to respond physiologically to photoperiod, one of the most common and error-free ways of initiating and terminating diapause in a tremendous number of insects.

Therefore, it is possible that the Dainty Sulphur has a critical preadaptation needed to evolve a true diapause as larva, pupa, or adult. Given the possession of this important preadaptation, the ubiquity of diapause among insects, and the fact that several other close relatives in the Coliadinae (such as *Colias*) have well-developed diapause capabilities in different life stages, it is possible that the midwestern segregate could evolve diapause in a relatively short period, as has apparently happened in several pest lepidopterans such as the Gypsy Moth (*Porthetria dispar*).

Unless the Dainty Sulphur develops a diapause, individuals making the long-distance migration northward are always doomed—and possibly have a lower fitness compared to those that do not migrate. However, the increased reproductive and survival capabilities of melanic forms confer three advantages that might facilitate the evolution of diapause and subsequent ecological range expansion by *Nathalis iole*.

First, greater reproductive capacity of melanic forms under cold conditions increases the likelihood that the genetic combination for winter diapause will arise among many possible genotype combinations, and then be selected for by the environment. Second, if diapause evolves, the extended flight activity of melanic fall forms and the early appearance and survival of melanic spring forms should ensure greater populations to survive the winter and to begin the summer brood production. Finally, cold-tolerant melanic forms would shorten the time the last brood spent in diapause, a time of high fatality from physical and biological factors.

Someday a population of Dainty Sulphur butterflies may develop a limited ability to diapause, perhaps in the southern portion of its present range where winters are short and consist only of intermittent cold spells. Once a limited ability to diapause takes place, selection should improve the trait so that the midwestern segregate becomes capable of surviving the more severe northern winters.

If diapause evolves, the Dainty Sulphur will have the final adaptation needed to complete a permanent ecological range expansion. This, of course, is an excellent pathway to speciation, so that some day we may document evolution in the act, through a prediction made long before the actual event. Such speciation events have obviously taken place in the complex sulphur genus *Colias*, and the Dainty Sulphur is in a similar physiological state as the presumed ancestor of its sulphur (*Colias*) relatives (Douglas and Grula 1978).

CHAPTER 6

Parasites, Parasitoids, Predators, and Defense

Every stage of a butterfly's metamorphosis is vulnerable to injury and death from any number and combination of physical and biological factors. Even the barnaclelike eggs, constructed to resist crushing and abrasion, fall prey to unseasonably cold temperatures, torrential downpours, and exposure to intense heat and sunlight. Furthermore, inclement weather may permit the invasion of pathogens such as molds and fungi that destroy the developing embryo within.

Biological factors such as minute parasitic wasps also take their toll of eggs. Downey (1957) records several minute parasites that attack the eggs of the Blue butterfly *Plebejus icarioides*. In some cases, more than one parasitic wasp larva inhabits a single butterfly egg. *Plebejus* eggs are about the size of a small pin head, so you can imagine how small the adult wasps must be! Unfortunately, the effect of egg parasitism on the population of any butterfly species is so poorly understood that virtually all information is anecdotal. In Downey's observation, up to 69 percent of a Utah clutch of eggs were parasitized, indicating that egg parasitism has the potential to significantly affect population densities, at least at the local level. In another study, however, White (1973) found only 15 percent of *Euphydryas editha bayensis* larvae parasitized.

There is simply not much an egg can do to protect itself. For the most part it is at the mercy of the weather, parasitic wasps, and a diverse assemblage of arthropod predators ranging from beetles to true bugs. The females of some Old World butterflies (e.g., *Nordmannia myrtale*), camouflage their eggs with anal hairs, but the effectiveness of this strategy is unknown (Nakamura 1976).

The larval stage probably suffers the highest rate of attrition. Here we have a vulnerable soft-bodied insect with limited movement, whose job it is to consume food as rapidly as possible and undergo periodic molts, each of which is an exercise in survival in itself. Caterpillars are attacked by fungi, bacteria, and viruses to the extent that whole populations may be decimated.

Nuclear polyhedrosis and *granulosis* viruses, so named for the peculiar crystallike inclusions formed within infected cells, are particularly devastating. Viruses literally putrify the innards, so that in death the caterpillar is strung out like a wet sock, sometimes hanging limp by a single proleg from its perch. Here they blacken and decay. Some cultures of butterflies, such as the Orange Sulphur (*Colias eurytheme*)—a common research animal—are highly susceptible to viruses that may be transmitted to the eggs via the contaminated ovaries of the female. A gentle washing of the eggs in a mild bleach solution for about sixty seconds effectively eliminates this problem. There are undoubtedly hundreds of species of bacteria and fungi that attack larvae occasionally, or with some degree of specificity, but again, virtually nothing is known about larval mortality from microorganisms under field conditions. Commercial preparations of the bacterium *Bacillus thuringiensis* are effective, however, against a broad spectrum of lepidopterous pests.

Arthropod parasites and predators of butterfly larvae are better known largely because they are more visible, hence more easily studied than pathogenic microbes. Virtually all arthropod predators—spiders, scorpions, praying mantids, true bugs, and predaceous wasps (they make "caterpillar hamburgers" for their larvae)—will attack and consume butterfly larvae if the opportunity presents itself.

Most of these predators do not have specific hosts, and they will attack almost anything smaller than themselves but large enough to meet their physical requirements for capture. Ants are particularly rapacious predators, especially in the tropics, and they will attack any stage. Larger predators include insectivorous birds and many species of small mammals such as rodents (Baker 1970). In subtropical and tropical areas, lizards are effective predators (Ehrlich 1982).

Larval parasites also include animals more aptly described as parasitoids because they are really internal predators that kill their hosts. Parasitoids include tachinid and dexid flies, as well as many species of minute hymenopterans, especially ichneumonid, chalcid, and braconid wasps. These tiny predators usually lay eggs in the caterpillar by penetrating the cuticle with a fine, needlelike ovipositor (e.g., Stamp 1984). Flies typically lay eggs outside the cuticle and the larvae burrow through the cuticle after hatching. In either case, the parasitoid larvae consume the butterfly larvae from within, attacking fat bodies, muscles, and later the gut and nervous system, but avoiding vital organs until the final days of metamorphosis. The host larva perishes before or during pupation, as the parasitoid pupates within the larva or on its surface.

Because few butterflies are of economic importance, actuarial tables of survivorship at different stages of metamorphosis are virtually nonexistent. Larvae suffer high mortality certainly, from extreme weather, improper choice of food plant by the ovipositing female, parasites, predators, parasitoids, and minute pathogenic microorganisms, but how important each of these population-regulating factors is by itself or in combination remains a deep, dark secret. Suffice it to say that relatively few larvae survive to pupate.

There are a number of interesting strategies by which larvae passively or actively defend themselves from predators and parasitoids. For example, swallowtail (papilionid) larvae—even the larvae of the primitive swallowtail genus *Baronia*—have a fleshy Y-shaped structure hidden behind the head called the *osmeterium*. The osmeterium is everted when the caterpillar is disturbed or threatened, releasing foul-smelling odors. The odors are generally characteristic of the aromatic compounds found within the leaves of the larval food plant and may serve to ward off or "disgust" potential predators. If pinched, a swallowtail larva everts the osmeterium, rears the head and thoracic region, and thrashes about wildly. This action is certainly an effective defense strategy, and many vertebrates, including humans, recoil at the sight of a thrashing green head equipped with repugnant orange horns.

Other species of papilionids such as the Spicebush Swallowtail (*Papilio troilus*) bear false eyespots in later instars that make them appear to be larger, more dangerous animals. In contrast, the newly emerged larvae are black with a conspicuous central white patch or saddle that makes them resemble bird droppings, or leaf aberrations. Many nymphalid larvae are covered with sharp spines and irritating hairs that presumably help thwart oviposition by parasites and parasitoids, and possibly discourage some predators. If disturbed, these caterpillars curl up into a ball and fall to the ground, where they remain motionless.

The common Baltimore Checkerspot butterfly, *Euphydryas phaeton*, has a variety of escape and defensive behaviors that vary among different instars. The caterpillars can effectively ward off the tiny parasitoid wasp, *Apanteles euphydryidis*, which exhibits attack behaviors geared to the different prediapause instars of the butterfly. However, parasitism was not a major factor in most population fluctuations of *Euphydryas* species, either among different localities or in different years. It is extremely interesting, however, that these minute parasitoids probably exhibit various search-and-attack behaviors that depend on the changing behaviors of the host caterpillars (Stamp 1984).

By and large, larvae use passive defenses, especially camouflage or *crypsis*. The colors and patterns of some larvae match the food plant with a precision that is inspirational. Bizarre larvae exist that resemble leaf blotches and veins, flower petals, and even sticks of wood. It would appear that such well-camouflaged larvae would be almost impossible to detect, but there are no well-designed studies to test this assumption.

More persuasive passive lines of defense include storing or sequestering poisonous plant by-products into the cuticle or haemolymph (e.g., Brower and Brower 1964a, 1964b). For example, many ithomiid, heliconiid, and danaid species sequester poisonous alkaloids and other compounds from their larval food plants. The best-known example is that of the Monarch whose larvae sequester *cardiac glycosides* (cardenolides) from their milkweed food plants. These toxins act as vertebrate heart toxins and emetics. The foul-tasting Monarch caterpillars are usually camouflaged, but once seen, their pattern is quite conspicuous and memorable—as if to advertise their poisonous properties once discovered. After only one emetic bout, most predators learn quickly to avoid anything that even remotely resembles the terrible-tasting caterpillars (Brower, Ryerson, Coppinger, and Glazier 1968).

Brower and his colleagues discovered a "palatability spectrum" in Monarchs (Brower, Seiber, Nelson, Lynch, and Tuskes 1982) that was related to the amount (or lack) of cardiac glycosides in the larval food plant. For example, the larvae reared on some milkweeds (*Asclepias*) are six times as emetic as those reared on other species (Brower, Ryerson, Coppinger, and Glazier 1968). Thus, the cardiac glycosides that Monarch butterflies sequester as larvae differ greatly in their emetic potency.

The cardenolides are also concentrated to different degrees in different parts of the body as well as in the two sexes. The cardiac glycosides in the abdomen have a higher emetic potency than those in the rest of the body, but wings have the highest concentrations of all, and males have lesser concentrations than females (Brower and Glazier 1975). The high concentration in the wings may maximize the deterrent efficiency of the cardenolides, as butterflies are frequently caught by their wings and then released once the bird gets a bitter taste of the toxin. Each Monarch contains more cardiac glycoside than the equivalent dose of digitoxin initially administered to adult humans suffering from acute congestive heart failure (Brower and Glazier 1975).

Even so, some birds such as black-backed orioles (*Icterus abeillei*) and black-headed grosbeaks (*Pheucticus melanocephalus*) can overcome the Monarchs' toxic defenses, and each year flocks of these birds eat hundreds of thousands of Monarchs as they roost in their Mexican overwintering sites (Calvert, Hedrick, and Brower 1979). In some roosts these birds are responsible for 60 percent of the butterfly mortality. Orioles evade the cardenolide poisoning by stripping out the thoracic muscle and abdominal contents without eating the cardenolide-laced cuticle. Other birds may have physiological ways of detoxifying these poisons (Brower and Glazier 1975).

Some have suggested that Monarchs evolved their unpalatability through *kin selection* (Hiam 1982), in which individuals share the predation load by sharing the same genes and the capacity to sequester and advertise the presence of toxins. If members of the same species of poisonous butterfly live in the same areas (e.g., overwintering roosts of Monarchs, or night roosts of *Heliconius*) then predators that sample one butterfly will become sick and quickly learn about their emetic properties. Although one individual has been sacrificed, the traits required for unpalatability will be passed on through the genes of its relatives—hence, kin selection.

Recently Brower and his colleagues have discovered variations in cardenolide content between plants and even within a single plant throughout the growing season (Nelson, Seiber, and Brower 1981). They also developed a "fingerprinting" method that allows them to analyze the type and amount of cardenolides in a Monarch butterfly. Such a chemical profile will permit them to identify the species of milkweed that the Monarch fed on as a larva and perhaps even place its geographical birth site (Brower, Seiber, Nelson, Lynch, and Tuskes 1982).

Cardenolides have also been discovered in North American *Erysimum* (Cruciferae). These cardenolides may serve as chemical defenses, and they may explain the plant's toxicity to native pierid butterflies in Rodman and Chew's (1980) study. Until recently, glucosinolates (mustard oils) were thought to be the standard crucifer toxin (Rodman, Brower, and Frey 1982); now we know that both glucosinolates and cardenolides may co-occur in plants. Rausher (1982) has found that populations of *Euphydryas editha* grow best on their native food plant, and that this maximal growth is not due to differences in feeding behavior, but apparently to population differences in digestive physiology. In time, these adaptations at the population level could permit newly founded populations to become a new species, as oviposition behavior, larval feeding behavior, and larval digestive behaviors are altered (see also chap. 8).

In one respect, the evolution of plant toxins and the continued evolution of the herbivorous predators of those plants to overcome those toxins has been viewed as a coevolutionary arms race (e.g., Berenbaum and Feeny 1981; Berenbaum, 1983). For example, although xanthotoxin (a linear furanocoumarin found in plants of the carrot family Umbelliferae) is not very toxic to the Black Swallowtail (*Papilio polyxenes asterius*), angelicin (an angular furanocoumarin found in a relatively few advanced tribes of umbellifers) reduces growth rate and fecundity. The presence of angular furanocoumarins may have been due to predation by the Black Swallowtail. However, even plants with the angular furanocoumarins are not immune to counteradaptation by insects, as the Short-Tailed Swallowtail (*Papilio brevicauda*) feeds almost exclusively on umbellifers with the angular form (Berenbaum and Feeny 1981).

The linear furanocoumarins (also called psoralens) are toxins that chemically alter the genetic material, DNA, when the psoralens are activated by long-wave ultraviolet light. Black Swallowtail larvae can apparently detoxify the psoralen xanthotoxin by special enzymes in the body and gut tissue. The ability to detoxify these linear furanocoumarins has evolved independently in other lepidopterous insects such as the Fall Armyworm (*Spodoptera frugiperda*) (Ivie, Bull, Beier, Pryor, and Oertli 1983).

Thus we have a coevolutionary war going on between plants and butterflies. When a new secondary plant compound arises by mutation, it may by chance affect the suitability of the plant as food for its predators. Plants with these new compounds may escape predation from insect herbivores and undergo *speciation*—radiating into many different species. However, predators may evolve a resistance to the new (formerly toxic) secondary plant compounds, and even sequester these compounds as toxins in their own defense (Berenbaum 1983). In this way, Monarchs and other danaid butterflies may have helped cause the evolutionary radiation of milkweeds (*Asclepiadaceae*); likewise with *Heliconius* butterflies and the incredible diversity of passionflower species (Passifloraceae; see also chap. 8).

Rather than chemical defenses, the larvae of Milbert's Tortoise Shell (*Nymphalis milberti*) and the Painted Lady (*Vanessa cardui*) construct webs over the growing sections of their food plant. Communal webs of Milbert's Tortoise Shell may harbor hundreds of spiny caterpillars, while the Painted Lady caterpillars typically build individual webs by weaving two or three leaves together and then skeletonizing them. Webs hide the caterpillars from the searching eyes of arthropod and avian predators alike, but they also serve to moderate the thermal environment within, possibly making it more conducive to growth.

Larvae may also be communal yet not construct webs. For example, the larvae of the small nymphalid butterfly *Chlosyne gorgone* are gregarious during the early instars but do not construct a communal web. Two possible advantages of gregarious behavior are that communal caterpillars somehow deter the attacks of predators and parasitoids, and perhaps also allow young larvae to overcome plant mechanical defenses such as hairs and trichomes more effectively than would solitary larvae (Rathcke and Poole 1975).

The larvae of *Chlosyne lacinia crocale* not only feed gregariously under a silk web, but hatch synchronously from the eggs. The communal early instars may maximize eating time and minimize energy expended by being gregarious. However, later instars tend to wander during the afternoon and aggregate at night. Perhaps the dispersal in later instars reduces the chance for disease, parasitism, and predation since larger larvae are more conspicuous. Thus it would appear that the costs and benefits of group living change with age (Stamp 1977).

Perhaps the most interesting line of defense is the mutually beneficial relationship that has evolved between ants and lycaenid butterflies. Many species of lycaenids sport glands on the abdomen, some eversible, and all of which produce either odors attractive (or repellent) to ants, or a sugary *honey dew* secretion that the ants extract for food. As with their aphid hosts, the ants stroke the lycaenid larvae with their antennae, apparently stimulating the production of more honey dew. The highly social ants are opportunistic, and protect the larvae from parasitoids and predators. They may also move some larvae from plant to plant (Downey 1962b) like humans moving cows from pasture to pasture.

The hairstreak *Harkenclenus titus,* and the blues *Glaucopsyche lygdamus* and *Celastrina argiolus pseudargiolus*—all known to be *myrmecophilous* or "ant-loving" (perhaps it should be lycaenophilous or "lycaenid-loving" since the ants are doing the stroking)— have recently had some of their attendant ants identified. In southern Michigan these include *Formica subsericea* and *Camponotas nearcticus* (Harvey and Webb 1980), both also known as attendants of bracken fern in Michigan, a plant that provides ants with axillary nectaries that ooze sugar and free amino acids (Douglas 1983).

Downey (1965b) reports thrips as well as numerous species of ant attendants associated with *Plebejus icarioides.* This suggests that ants will tend, and probably protect, any resource that exudes a sugary solution, whether it be fern, aphids, or lycaenid larvae. In fact, it is very likely that each lycaenid species may be attended by different species of ants in different areas of its geographic range. Thus, the mutualistic relationship between ants and lycaenid larvae is undoubtedly opportunistic to some degree, involving different partners in different areas.

Several researchers have proposed that food plant choice by lycaenid butterflies is affected by what is termed "enemy-free space" (Price, Bouton, Gross, McPherson, Thompson, and Weis 1980). Because the lycaenid-ant relationship is hypothesized to have developed in part as a protective measure against larval and pupal enemies, including parasitoids and predaceous ants, lycaenid adults might not choose necessarily the most beneficial food plant, but one in which attendant ants are found (Atsatt 1981a). If this is true, then unlike other groups of butterflies that are generally restricted to perhaps several related families of plants, the ant-adapted lycaenids should hypothetically be more diversified in their tastes—more able (rather than willing) to try novel food plants—as long as ant defenders are present.

And this appears to be the case, for although most lycaenids consume flowering plants (including leaves, flowers, and fruits), they also have diversified onto fungi, lichens, cycads, ferns, conifers, and even ant larvae and homopterans such as scale insects! Furthermore, six of ten of the lycaenid subfamilies are tended by ants, and Atsatt (1981a) has determined that generic diversity increases tremendously when ants are present in the larval development period. Thus, there are 368 genera of lycaenids associated with ants, and only 24 genera of lycaenids without ant attendants. Finally, within subgroups of lycaenids, shifts in food plant preference are unpredictable and commonplace (Downey 1962a). Thus, as Atsatt points out, you could not accurately guess the food plant of an unclassified lycaenid butterfly even if you knew the lycaenid species most closely related to it.

Perhaps the honey dew secretion of lycaenid larvae not only attracts ant protectors, but prevents the ants from attacking the larvae. This is precisely what homopterans (primarily aphids) do with their ant attendants. In this scenario, lycaenid larvae would develop an enemy-free space each time a population successfully communicated chemically with ants associated with its food plants. Many species of lycaenids have evolved peculiar morphological and behavioral traits that reflect coevolutionary adaptations.

For example, ovipositing females can locate ants on potential food plants. The slug-shaped larvae that emerge have much thicker cuticles than normal (Malicky 1970) so that the ants can pick them up without penetration. In cross section, the ant-attended lycaenid cuticle is up to sixty times thicker than that of a nonlycaenid caterpillar of the same size. And of course, the larvae have a variety of nectar glands—some associated with eversible tentacles with distal setae that whir about (Malicky 1970). Most importantly, though, ants palpate lycaenid larvae with their antennae, striking different spots of the cuticle in different species, and Malicky (1970) has shown that these species-specific areas contain minute glands (perforated cupolas) believed to produce volatile substances, possibly amino acids (Atsatt, personal communication), that attract and pacify the investigating ants. The nectar gland on the dorsal posterior margin of the seventh or eighth abdominal segment may contain up to 18 percent sugar, and at least one free amino acid has been identified.

When the paired eversible tentacles are present, they are located just posterior to the nectar gland. When the larvae travel, the tentacles may dispense chemicals that cause the attendant ants to disperse (Claassens and Dickson 1977). The larvae are thought to follow the dispersing attendant ants to protected sites where they are sheltered when not feeding.

The last instar larvae of some lycaenid species are carried to the ant nests by attendant ants, where prior to pupation, they feed on ant larvae and pupae, still appeasing their attendants with ambrosia from the nectar gland. When these butterflies eclose after diapause, they are covered with superfluous scales that are easily rubbed off—and with good reason. The emerging adults are often attacked by the ants, but the scales become lodged in the ants' tarsi and mandibles, causing them to retreat. Thus, scales also serve as a dry lubricant, allowing the butterflies to emerge from ant nest holes relatively intact (Atsatt 1981a).

In some cases, even the selection of larval food plant by adult lycaenids is "ant-dependent"—that is, ants must be present before females will oviposit. Females that are actually touched by ants usually lay the largest number of eggs. For this reason, the lycaenid *Ogyris amaryllis* will elect *Amyema maidenii*, a nutritionally inferior food plant, over *Amyema preissii*, if attendant ants are present (Atsatt 1981b; Atsatt, personal communication). More eggs are laid with ants present and these eggs suffer lower parasitism rates than those laid on food plants without ants. Strangely, though, first and second instar larvae of *O. amaryllis* do not secrete honey dew. Somehow, apparently through the use of behavioral controlling pheromones, these first two instars manage to attract—and appease—their attendant ants. Larval nectar production may be timed so that more conspicuous, larger larvae are protected. Thus, in at least one case it can be shown that ants affect the survivorship of lycaenid larvae (Pierce and Mead 1981), and lycaenid adults of *O. amaryllis* may choose chemically less suitable host plants as long as ants are present to enhance survivorship.

The *O. amaryllis* story may represent a common strategy whereby lycaenid species with ant associates maintain a larger enemy-free space, and hence have the opportunity to take chances in the evolutionary sense with host plants chemically less desirable than the preferred hosts. Such a strategy could be responsible for the tremendous adaptive radiation observed within the Lycaenidae and their host plants. In turn, such lycaenid–ant–host plant associations could be responsible for rapid speciation events since the probability for reproductive isolation is greatly increased.

Such complicated ecological relationships may also be more susceptible to disruption. For example, the destruction of habitat containing the two ant mutualists of the beautiful European Blue, *Maculinea arion*, is thought to have contributed to its recent extinction in England (Ratcliffe 1979). The caterpillars are remarkably camouflaged to match the flowers of their larval food plant, wild thyme, for the first three instars. Their food substrate in later instars prior to larval hibernation in the ant host nest is ant brood.

Atsatt suggests that there are two basic lycaenid "strategies." In species that feed on early successional herbaceous plants and/or ephemeral parts such as buds, flowers, and fruits, larvae are not consistently associated with ants, and therefore less likely to have evolved in a coevolutionary mutualism with them. Their defenses are camouflage and the mimicry of unpalatable insects or objects. The second strategy involves lycaenids that utilize more "apparent" plants—that is, predictable and long-lived species such as late successional woody plants. These lycaenid larvae in turn become "apparent" to another trophic level involving ants. Because ants can provide an adequate defense against parasitoids and predators, these lycaenids have the opportunity to become more catholic in their food plant choices—as long as ants are present.

Thus, the second strategy sets up and encourages host plant shifts. Perhaps this is why biochemical similarity between host plants plays a lesser role in oviposition selection by adults. They may in effect be guided to a reasonable, but perhaps not best, food plant simply because suitable ant attendants are found there. Genetically and behaviorally, changes in the selection of larval food plants by adult butterflies are theoretically straightforward. Such a strategy also helps explain why the males of some lycaenid species are highly territorial—that is, they defend a flight space or basking perch. For example, males of the lycaenid *O. amaryllis* constantly patrol individual ant-occupied mistletoes (Atsatt 1981b).

The defense strategies of butterfly pupae are not nearly so interesting as those of larvae and adults. Lycaenid pupae in ant nests obviously receive protection not only from predators and parasitoids, but also from extreme weather. Subterranean climates are scarcely affected by too much rain, not enough rain, extreme cold, or extreme heat. In addition, some lycaenid pupae stridulate, which also has been hypothesized to provide a defensive function (see chap. 2).

The larvae of some butterfly species, such as the Mourning Cloak (*Nymphalis antiopa*) pupate gregariously, sometimes forming clusters of hundreds of chrysalids. Aggregations of pupae are also common in tropical species of butterflies, especially in the Nymphalidae and Pieridae. Gregariousness in pupae seems to be a logical outcome of gregariousness as larvae. There are several advantages to larval gregariousness, but of what evolutionary significance is pupal gregariousness?

Possibly, this behavior ensures quick mating, especially if adult eclosion is synchronous. Then again, pupal aggregations might possibly serve some sort of defense mechanism that varies from species to species (Young 1980). If one pupa of the Mourning Cloak butterfly is disturbed, it thrashes back and forth, slamming the head region into its support—typically a branch, log, or piece of bark. Since the spacing is so close, its thrashing cannot help but disturb the other pupae, which instantly respond with thrashing of their own. The motion and noise of fifty to one hundred chrysalids can be considerable, and the unexpected commotion is enough to startle an unsuspecting person, and certainly must be sufficient to scare away small predators.

Still other pupae appear grotesquely misshapen, resembling dried leaves, inanimate buds or twig stumps, and even snake heads—all in the name of camouflage. The pupae of *Dynastor darius*, a nymphalid butterfly from the Neotropics, look remarkably like the head of a snake with its mottled color pattern, distinctive shape, and snakelike eyes (Aiello and Silberglied 1978a). Aiello and Silberglied suggest that predators are so surprised by the snakelike appearance that they turn and escape, especially when the pupa thrashes madly back and forth.

Some of the most exciting defense mechanisms are found in adult butterflies, which, like larvae, are attacked by many of the same predators. Mites have been found on butterfly larvae, but typically they are true parasites of adults, sucking haemolymph from the base of wing veins and from membranous areas of the body. Although they are more commonly found on moths, Treat (1975) reports a number of mite species from nearly all major families of butterflies. Most of these mites belong to the Trombidiformes, the so-called red velvet mites that attach to the adult, engorge themselves to the bursting point, then drop off to complete their metamorphosis—a physiological feat that rivals that of the butterfly.

The eyeless mite *Hericia georgei* belongs to a genus whose members typically inhabit wet environments, and are commonly found at tree injury sites that exude sap (Treat 1975). This may explain why nymphalids like the Mourning Cloak and Red Admiral are sometimes infested, since oozing sap is one of their dietary delicacies in the early spring and late fall. The Trombidiformes are closely related to those decidedly unpleasant acarines, the trombiculids or chigger mites. Anyone who has collected butterflies below the forty-fifth parallel knows what misery the immatures of these mites can cause. Unless a butterfly is heavily infested, mites probably inflict little damage, although some researchers have reported that butterfly flight is weakened when large numbers are present.

Adult butterflies are especially vulnerable to injury and attack during and immediately after eclosion from the pupa. But once the wings have expanded and hardened, even avian predators have a tough time getting any more than a mouthful of scales. Except for the most agile of birds, butterflies can dodge and outmaneuver most vertebrate predators, including frogs, toads, lizards, and, of course, lepidopterists. Their downfall comes during mating, oviposition, basking, or foraging, when the adults are relatively sedentary and easily caught by a scissorlike beak—or between a forefinger and thumb.

Vertebrates are not the only successful predators of butterflies. A significant number of invertebrate predators—ants, beetles, dragonflies, true bugs, spiders, and scorpions for example—attack the adults. Immature crab spiders (Thomisidae) attack many species of butterflies. In particular, the genera *Misumenops, Misumenoides,* and *Misumena* commonly inhabit many species of goldenrod (*Solidago*)—ideal sites for ambushing butterflies (Jennings and Toliver 1976). Some crab spiders even absorb ultraviolet light in a pattern that matches the ultraviolet-absorbing pattern of the flowers where they lurk, waiting for passing insects (Hinton 1976). This pattern may conceal them both from butterflies and spider predators that are insensitive to ultraviolet wavelengths. However, Silberglied (1979) has pointed out that some ultraviolet-absorbing spiders occupy ultraviolet-reflecting flowers, and therefore should be quite visible in these wavelengths.

Larger species of dragonflies will also attack butterflies, especially if they happen to enter the territory of a patrolling adult. Dragonflies establish definite territories along streams or permanent bodies of water, and they will patrol and defend these territories from other species as well as from members of their own species (Manley 1971). True bugs (Order Hemiptera) are also effective predators. They typically hide in flowers, especially composites, awaiting visitations by butterflies. Pyle (1973) records a phymatid bug attacking *Boloria selene* in Washington state that had the misfortune of probing a flower in which several bugs were hidden.

Fales (1976) documents several additional accounts of attacks on butterflies by the ambush bug *Phymata fasciata,* and it is likely that predation on Lepidoptera, including butterflies, by ambush bugs is much more common than realized. In many cases, the butterflies are captured when the ambush bug seizes the butterfly's proboscis as it feeds, often mangling it in the process. As the butterfly is snatched close, the bug sinks its beak into the victim and liquifies the interior with digestive enzymes.

Although butterfly predation by birds seems uncommon and is rarely documented, attacks do occur under special circumstances. For example, even unpalatable Monarch butterflies, many of them with high concentrations of cardenolides, are attacked within their overwintering roosts (Brower, Calvert, Hedrick, and Christian 1977). Calvert and his coworkers report "extraordinarily heavy" mortality at five overwintering sites (see chap. 5) in Mexico where the forest floor was thickly carpeted in places (up to 770 individuals per square meter) by butterfly wings and pieces of bodies. Over 90 percent of these butterflies died from predation; nearly 70 percent had holes in the thorax, and over 50 percent were missing their abdomens (Calvert, Hedrick, and Brower 1979).

Three species of birds are responsible for the heavy predation: Scott's Oriole (*Icterus parisorum*), the black-backed Oriole (*Icterus abeillei*), and the black-headed grosbeak (*Pheucticus melanocephalus*). The orioles and grosbeaks forage among the dense roosts of mixed flocks containing twenty-five to thirty birds. The orioles peck the butterflies and rip out the abdominal innards, whereas the grosbeaks often snap off and eat only the abdomens.

Brower and his coworkers also noticed that in 37 percent of the attacks, Monarchs were hit, sometimes damaged, and then released. They hypothesize that these butterflies contained sufficient doses of cardenolides to make them unpalatable to the birds. In fact, Monarch butterflies exhibit a "palatability spectrum," and Brower and his colleagues (Brower, Ryerson, Coppinger, and Glazier 1968) had earlier determined that blue jays (*Cyanocitta cristata bromia*) can discriminate palatable from unpalatable butterflies by their cardenolide content. Novice jays that consume poisonous Monarchs become violently emetic. Only one such trial is required to train a blue jay for a long, long time (Brower and Glazier 1975).

At Site Alpha, 29 percent of the butterflies contain less than 10 micrograms of cardenolide (equivalent to 0.00035 ounce), which is an amount that is apparently tolerable by avian standards. Thus, these three bird species, and likely others, have penetrated the Monarch's chemical defense system, or at least can recognize that some Monarchs are palatable mimics of unpalatable Monarchs (Fink and Brower 1981).

Larger colonies possibly suffer less predation because the predators tend to forage on the periphery of the colony. Smaller colonies present more butterflies on the surface of the colony, and therefore experience potentially greater losses. Another factor that may limit colony size, at least in evolutionary terms, is volcanism, a geological activity endemic to the Central Transvolcanic Range of Mexico. Forest fires and smoke can and probably have destroyed entire colonies in the past. Thus, more overwintering sites increase the likelihood that the species will survive in the event of volcanic catastrophe and resulting fires (Brower 1977).

Predators that attack distasteful or poisonous species of butterflies quickly learn to associate the color and pattern of their wings with unpalatability. In the case of the Monarch, not all milkweeds are poisonous, and hence not all Monarchs are poisonous. Nonetheless, the palatable individuals receive some protection because they cannot be told visually from unpalatable ones. Thus, the palatable butterflies are, in a way, *automimics* of the unpalatable Monarchs. Other species of butterflies have taken advantage of this general unpalatability by evolving a color pattern mimicking that of the Monarch. One such butterfly is the Viceroy, *Limenitis archippus*, a palatable species whose larvae feed on willows, poplars, and related trees, but apparently gain no chemical protection from these food plants.

The palatable Viceroy mimics the unpalatable Monarch over most of the Monarch's range, with the exception of the southern tip of Florida, where the Florida Viceroy (*Limenitis archippus floridensis*) exists as a sharply defined geographic race (or *subspecies*) that is a rusty brown in color. Here the Monarch is rare, but another unpalatable cousin, the much darker Queen butterfly (*Danaus gilippus berenice*), serves as the model.

Similar subspecies of the Viceroy are found in central Texas and the southwestern United States where different models of the Queen butterfly occur. Thus, the wing color and pattern of the Viceroy vary geographically in accordance with the unpalatable model present. *L. a. archippus* inhabits the north and east, *L. a. floridensis* peninsular Florida, *L. a. watsoni* southern Texas, and *L. a. obsoleta* the southwestern United States. Each subspecies resembles a different unpalatable danaid model, and narrow zones of "blending" occur where the models switch over (Oliver 1972). In this mimicry complex, a palatable butterfly mimics the pattern of an unpalatable species wherever their geographic ranges overlap (Brower 1958a, 1958b).

Such *Batesian mimicry* complexes—named after Henry W. Bates, who first documented this phenomenon during the mid-nineteenth century in the Amazon Basin—are rather common, but especially in tropical areas. The temperate *Limenitis* complex is unusual in that all species are presumably palatable, yet there is a wide range of color and pattern differences. Also, in each species of mimetic *Limenitis* butterflies, both sexes have nearly identical phenotypes and hence both mimic an unpalatable model. This is termed *mono-*

morphic mimicry. In other Batesian complexes females of palatable species have different geographic or seasonal forms that mimic different unpalatable models, whereas males are relatively invariant— a type of *sex-limited mimicry.*

Actually, several species of *Limenitis* butterflies are involved in different mimicry complexes. Thus, while the Viceroy and its geographic subspecies are mimics of the Monarch and Queen, the Viceroy's congeneric, the Red-spotted Purple (*Limenitis arthemis astyanax*) mimics the unpalatable Pipevine Swallowtail butterfly (*Battus philenor*), wherever their ranges overlap (Brower 1958c). Other butterflies also presumably mimic the Pipevine Swallowtail, including the bluish females of the Diana Fritillary (*Speyeria Diana*) (males are typically "fritillary" orange); the females of the Tiger Swallowtail (*Papilio glaucus*) where their ranges overlap; both sexes of the Black Swallowtail (*Papilio polyxenes asterius*); the Spicebush Swallowtail (*Papilio troilus*); and possibly even the day-flying males of the giant Prometheus silk moth, *Callosamia promethea.*

Such a vast number of palatable mimics would seem to numerically swamp out the single unpalatable model, but even a protection of 5 percent is better than nothing. As long as unpalatable models fly in the same range as the mimics, the mimics hypothetically will gain some—as yet unmeasured—amount of protection. The case for Batesian mimicry in this group is strong, but largely unproven.

Just as the Viceroy changes phenotype in the extreme southern part of its United States range in order to mimic a closely related unpalatable species of the Monarch, there is a dramatic shift between two closely related *Limenitis* subspecies in the temperate zone where the northern range of the Pipevine Swallowtail stops. Here, the northern Banded Purple (*Limenitis arthemis arthemis*) and the Red-spotted Purple (*L. a. astyanax*) form a relatively narrow hybrid zone where interbreeding occurs. Hybridization results in the intergrade form *L. a. proserpina.*

The northern Banded Purple is *disruptively colored*—that is, patterned in such a way that the outline of the butterfly's wings are · broken up against a discontinuous background—while the more southern Red-spotted Purple probably mimics the Pipevine Swallowtail. Both the Viceroy and the Red-spotted Purple presumably evolved from a banded ancestor similar to the Banded Purple, be-cause most Palearctic *Limenitis* are banded, and this phenotype likely represents the ancestral color pattern (Platt 1975).

Platt has attempted to reconstruct the evolutionary history—transitions, phenotypes, and selective processes—that may have taken place as the two monomorphic mimics, the Viceroy and the Red-spotted Purple, diverged from a hypothetical banded ancestral form. He did this by rearing hybrid broods from the different *Limenitis* species and subspecies, observing interfertility and the inheri-tance and segregation of colors and patterns. The crosses between the Red-spotted Purple and the Banded Purple produced "blended" forms like the naturally occurring hybrid, *L. a. proserpina*. When hybrids were backcrossed to their parents, these traits were segre-gated again. Apparently, the major genes responsible for the phe-notypic changes are unlinked—that is, the genes are found on differ-ent chromosomes and thus segregate independently of one another during the formation of sex cells.

The possible evolutionary sequence constructed by Platt leads not only to speciation in *Limenitis* butterflies, but to radical diver-gence in color pattern. Both mimetic species and disruptively colored species are "fit" in the evolutionary sense. The phenotypic divergence was accompanied by behavioral divergence. We find that the Viceroy is a species of open fields whereas the Red-spotted Pur-ple and Banded Purple are riparian, woodland, or forest-edge species. Platt (1975) contends that the driving force in this speciation event was differential predation by visually oriented vertebrate predators such as birds.

There are many other examples of Batesian mimicry in the liter-ature, including butterfly species that mimic unpalatable day-flying moths, and mimicry complexes in the Neotropics in which day-flying moths have converged on a common "warning" color pattern presented by certain species of *Heliconius* butterflies. Generally, Batesian mimicry requires that the model be more common than its mimics, but whether or not this is always the case is not known nor has it been rigorously tested in the field.

There are many unexplained phenomena and questions surround-ing Batesian mimicry. For example, why do both sexes of some species mimic unpalatable models (monomorphic mimicry) while only one, typically the female, of other mimetic species do so (sex-limited mimicry)? Only females of the Tiger Swallowtail are melanic in geographic areas that overlap the distribution of the Pipevine Swallowtail model—are females more valuable than males in this species than in others? Or does this simply reflect a mode of mimicry inheritance involved that may differ between species? Where the Pipevine Swallowtail is absent, such as in the northern temperate states, the female Tiger Swallowtail is the typical yellow-and-black banded form.

Both male and female Red-spotted Purple butterflies are presumably mimics of the Pipevine Swallowtail. Likewise, both sexes of the Viceroy mimic danaid models. Does this mean that sexual dimorphism is not available to *Limenitis* butterflies? Obviously, once a form occurs that mimics any unpalatable model, that mimicking form should have greater survival value, and selection will increase its fitness value. The rationale is that any behavior or wing pattern that can be used to identify palatable mimics will be selected against by predators, thereby refining the mimicry.

Another form of mimicry, which may exist separate from or be mixed in with Batesian mimicry complexes, is termed *Müllerian mimicry* after Fritz Müller who, like Bates before him, first found evidence for mimicry in the Amazon Basin. A Müllerian mimicry complex consists entirely of unpalatable species whose ranges overlap temporally and geographically. All the species involved have converged on a common wing pattern and sometimes a common flight behavior as well.

In Müllerian complexes, each species is both a model and a mimic. Since all are distasteful, the convergence on a common pattern tends to reinforce the mimicry. There are no weak links like the palatable mimics in Batesian mimicry. If a predator samples any member of Müllerian complex, that butterfly should be rejected and all others resembling it should henceforth be safe from that predator. By contrast, novice predators should be statistically luckier with a Batesian mimicry complex: they may attack palatable mimics several times before being stunned by an inedible model.

A Müllerian mimicry complex is economical in the evolutionary sense—a common, well-advertized color pattern serves to reinforce the bad-tasting image of all members in the complex. The more species joining the complex, the more efficient its operation should be, and fewer losses should be sustained by each species in the complex. A Batesian mimic, however, is basically a freeloader—even, for example, the automimicking Monarchs that lack cardenolides. The Müllerian mimicry complex is a mutual defensive agreement struck between species, each doing its part in protecting the colors of the complex in the evolutionary sense.

There are numerous examples of Müllerian mimicry, many of which are found in the Neotropics (e.g., Brower, Brower, and Collins 1963). Even ultraviolet reflecting patterns may be shared in these complexes (Silberglied 1979). However, as with Batesian mimicry, the effectiveness of Müllerian mimicry complexes under field conditions is largely untested. One study by Benson (1972), however, does indicate that Müllerian mimicry involving the unpalatable butterfly *Heliconius erato* indeed operates to protect the member species from predation.

H. erato and its various races flies with another similarly patterned longwing congeneric, *H. melpomene*, which occupies the same geographic range and habitat. Benson altered the pattern of *H. erato* butterflies to a unique nonmimicking pattern. He then returned these individuals to natural populations and noted evidence of attempted predation, such as beak marks on the wings. The experimental design allowed the release of treated control butterflies whose pattern was unaltered, and treated experimental butterflies with altered wing patterns.

The experimental butterflies with altered wing patterns disappeared from the population at a faster rate and showed more evidence of vertebrate attack—chips, tears, gouges—than the control group with unaltered wing patterns. The results indicate that natural selection was operating to maintain the monomorphic phenotype in the mimicry complex, thus supporting the Müllerian mimicry hypothesis. Any naturally occurring forms distinguishable from the unpalatable phenotype will likely be selected against, whereas those closely matching the common pattern already shared by the majority of the species in the complex will survive longer and likely reproduce more successfully.

Perhaps the most amazing mixture of Müllerian and Batesian mimics was uncovered by Papageorgis (1975). She found that three or more distinct Müllerian mimicry complexes can exist in the same geographic area—contrary to the traditional interpretation of the Müllerian theory—as long as these complexes were stratified within the tropical rain forest canopy. Each complex converges on a different warning color pattern and occupies a different flight zone within the forest. Apparently, the color patterns have evolved independently of thermoregulatory considerations.

The mimicry rings (involving both Batesian and Müllerian mimics, moths as well as butterflies), are incredibly complex. In each of three Peruvian locations studied by Papageorgis, there were up to five distinct mimicry rings. The "transparent complex," containing mostly ghostlike moths and ithomiid butterflies with transparent wings outlined by black veins, fly below six feet. The "tiger complex" flies above them to heights of twenty feet and consists of yellow, brown, black, and orange-striped butterflies mostly in the ithomiid, heliconiid, and danaid families, although some pierids, nymphalids, and day-flying moths are also present. The "red complex" flies above the tiger complex between twenty and forty feet, and is dominated by the longwing butterflies *Heliconius erato* and *H. melpomene*. These two species have different geographic races with different patterns throughout their sympatric distribution, but in each locality both species have converged on a common pattern (Turner 1971, 1976). The red complex thus varies somewhat in pattern and species composition, depending upon the location.

A Müllerian "blue complex" of *Heliconius* butterflies occupies the upper canopy above forty feet, and a Müllerian "orange complex," also consisting of heliconiines, typically flies above the forest canopy. So fantastic are the color and pattern convergences within each complex that only pictures can effectively illustrate their complexity. It is clear that each complex is not only vertically stratified, hence largely isolated from other such complexes, but that closely related species have diverged greatly in pattern to join different complexes, much as members of the genus *Limenitis* have done in temperate North America.

Papageorgis suggests that the patterns of the complexes may have evolved in response to different light intensities received within a given level—hence flight range—within the forest canopy. The transparent species seem to dissolve into their background, and the blues appear to be disruptively colored at their flight level. Thus, the color patterns of Müllerian mimics need not be bright and flashy to be effective. They can be cryptically patterned, but they must be easily recognized by predators once located. Papageorgis (1975) concludes that it is the changing background of light and shadows within the forest canopy that has resulted in the differently patterned mimicry complexes, and the phenotypic divergence between closely related species.

Thus, camouflage and warning coloration are not necessarily mutually exclusive: butterflies may be both cryptically colored and distasteful. But mimicry is probably not as common a defense strategy as is camouflage, and adult butterflies may be camouflaged in a multitude of ways. Transparent species such as those found in the Acraeidae and Ithomiidae disappear before one's eyes—the transparency is due to the absence of scales, a scarcity of scales, or scales that are not fully expanded. In effect, the transparent species are matching not so much leaf shape or pattern, but patches of light and darkness common to their habitats.

Most butterflies have cryptically colored or patterned undersides. Many temperate nymphalids, for example the Red Admiral (*Vanessa atalanta*) and the Mourning Cloak (*Nymphalis antiopa*), match their resting substrates with remarkable precision. Their Indo-Malayan counterpart, the "dead leaf" nymphalid butterflies (*Kallima* spp.) have short tails and a midventral black stripe on the undersides of the wings that makes them look exactly like a dead leaf when the wings are folded. Reports of butterflies minimizing their shadow by tilting toward or away from the sun, a behavior especially prevalent in arctic and alpine species, also should take into consideration the thermoregulatory as well as the cryptic significance of such posturing.

Several pierid genera (*Pontia* and *Euchloe* for example), are distinguished by dark "green-colored" veins and patches on the ventral surface of the hind wings that effectively conceal them while at rest within vegetation (Bowden 1979). But depending on the background, virtually any color pattern can be cryptic. The light undersides of temperate *Colias* are difficult to see in a field of goldenrod, and the angle-wing butterflies in the genus *Polygonia* are almost impossible to detect when basking on fallen leaves.

Butterfly wings may also be disruptively patterned. The highly contrasting colors—usually bars or complete stripes—break up the shape of the butterfly and make it seem to disappear against a discontinuous background of light and shadows. The Palearctic Banded Purples (*Limenitis*) and their relatives are said to be disruptively colored because their broad white bands dissect the black ground color of the wings. This seems like a plausible hypothesis, yet the obliteration of the highly contrasting wing stripes in the Neotropical *Anartia fatima*—the tropical counterpart of our temperate *Limenitis* butterflies—did not affect survival or increase wing damage (hence presumed predation attempts) in a natural population over a five-month period (Silberglied, Aiello, and Windsor 1980).

Thus, there is no solid evidence that so-called disruptive colors are of adaptive value as protective coloration, or that these patterns necessarily function the way we hypothesize. Again, a great deal more evidence is necessary before the many hypotheses concerning wing coloration can be verified or negated. Yet many hypotheses, such as that of disruptive coloration, seem so logical—at least to us lepidopterists. Perhaps, though, butterfly predators do not exercise the same logic.

Some butterflies have cryptically colored undersides and brilliant, often iridescent upper sides. In the male *Morpho* butterflies, brilliantly flashing wings are seen briefly on the downstroke, and disappear on the upstroke. It is extremely difficult to follow such an on-and-off electric butterfly in tropical forests where shadows and lights play such an important role in crypsis. By contrast, unpalatable species such as the Monarch are not only gaudily colored in orange and black, but their patterns are prominently displayed during basking, foraging, and gliding flight. But not all distasteful butterflies need to be gaudily colored in order to advertise their unpalatability, as we have seen in the mimicry complexes described by Papageorgis.

There are also colors and patterns that serve to startle: lifelike eyespots and rows of smaller ocelli, or tiny eyes. Some species expose these eyespots when disturbed from rest, reportedly startling the intruder and frightening it away. Logically, this would appear to be the case, but this hypothesis also has not been tested rigorously in the field. The owl butterflies in the genus *Caligo* have some of the most vivid and remarkable eyespots, yet the butterfly does not rest upside down as often depicted, and so the stated "startle effect" on predators is simply not known.

It is known, however, that eyelike ocelli and eyespots at the wing margins serve to draw a predator's interest away from vital body parts, allowing the butterfly to escape without serious injury. Damage to eyespots, particularly to those in the anal area next to minute tails (most common in lycaenids and papilionids) is usually done by lizards and birds. In the case of many lycaenid butterflies, the hairstreaks and blues in particular, the minute tails and tiny eyes supposedly create a "false head" (e.g., Larsen 1982), directing attack toward less vulnerable posterior areas and away from the true head (Robbins 1980). The impression of a fake head is even more convincing when these lycaenids rub their hind wings together as they are folded dorsally over the back, making the fine, white-tipped tails appear very much like palpating antennae.

Robbins tested the false head hypothesis using the Neotropical lycaenid *Arawacus aetolus*, which is commonly cited to illustrate the false head hypothesis. Males of this species, like many temperate hairstreaks (e.g., *Satyrium calanus*) occupy basking and perching territories from which they dart out to chase available females, and also to ward off roaming males from their flight space. Robbins found evidence to show that the anal area of *A. aetolus* is frequently damaged (beak marks) as one would expect from avian predators— even though hind wing movements are sporadic and unpredictable. Robbins also observed that predators differentially attack the various wing patterns prevalent among lycaenids, and thus predation pressure influences the evolution of these patterns (Robbins 1981).

The foregoing discussion concerning the predators, parasitoids, and defense mechanisms of butterflies represents only a small fraction of the actual number of cases studied. And although many logical hypotheses exist to explain the colors and patterns of butterflies, particularly for defensive reasons, few have been tested rigorously under field conditions using a variety of butterfly species. Testing these hypotheses will take years, utilizing many different angles of experimental attack, but most importantly we should not lose sight of the fact that human logic may have no bearing on what is really happening in nature.

CHAPTER 7

From Sex to Speciation

Even the largest butterflies are small in comparison to most vertebrates. Somehow, though, male must meet female, and each must decide quickly if the other is an appropriate mate. Certain strategies have evolved to solve these problems. Generally speaking, the males do the seeking and the females do the selecting. Unfair as that may sound to some of us, the same strategy is used by the overwhelming majority of sexually reproducing species.

Males employ two basic mate-seeking behaviors: perching and searching. Perching species pick strategic sites from which to observe passing butterflies. Sometimes these perches are located near appropriate larval food plants; other times they are sites where visibility is unobstructed. Many perching species tend to be pugnacious, even territorial in a loose sense of the word, darting out and investigating any object that moves with a certain velocity across their visual field—even falling leaves, birds of prey, and lepidopterists.

Pugnacious males such as those of the Hackberry butterfly (*Asterocampa leilia*) may be residential in a given area for two weeks or more—often choosing the same perching site or sites (Austin 1977). Perching sites are probably chosen on the basis of more than one criterion. Unobstructed vision, the amount of sunlight, and protection from wind are likely to be important. The territory is defended to the extent that nonfemales or nonspecific intruders are chased away to the outer boundaries of the range claimed by the butterfly.

Males of some species actively search for habitats suitable for females, particularly larval food plants and certain species of flowers. Males may seek higher elevations—hills, outcrops, and even tall trees—that they patrol in a slow, gliding fashion. This hill-topping behavior probably allows searching males to make themselves visible to females and vice versa (Shields 1967).

In both perching and searching species potential mates are nearly always located by visual cues based on wing pattern and color in the visible region. However, many species are also exceptionally sensitive to long wavelengths (far red) due to the presence of red-absorbing pigments such as rhodopsin and retinal (Bernard 1979). These pigments suggest that the red, orange, and yellow markings of butterflies may be important for inter- and intraspecific sexual communication, as well as serve as warning colors in some species.

For example, Crane (1955) altered the color pattern of *Heliconius* butterflies and changed their mating success, while Silberglied, Aiello, and Lamas (1980) altered the pattern of *Anartia* butterflies and also affected mating success. This would suggest that color pattern is important to reproductive success. Shapiro (1983) also tested the hypothesis of visual species recognition in the Buckeye butterflies (*Precis coenia* and *Precis nigrosuffusa*) by manipulating their phenotypes (see also Hafernik 1983). He found that male *P. coenia* apparently could discriminate against bandless females, *P. nigrosuffusa* females, and phenocopies of *P. nigrosuffusa* (which lack bands) induced by chilling pupae of *P. coenia*. Despite these studies, Silberglied (1984) noted that there is little solid evidence to support Darwin's argument that selection for color pattern by male or female butterflies has been a potent force in the evolution of butterfly coloration.

Males' preferences probably do not greatly influence female wing color and pattern: females generally have low pattern diversity among closely related species, and sex-limited mimicry complexes (see chap. 6) and female polymorphism all support this argument. In addition, males usually court females one at a time, male reproductive success is limited by the number of receptive females encountered, and males can mate many times. Also, only ultraviolet visual signals on the males' wings of some species have been shown to affect female behavior (Silberglied and Taylor 1973).

Silberglied (1984) suggested that the brilliant color pattern of male butterflies serve in male-male interactions, since the recognition of competing males and the advertisement of their own male sex would be advantageous for males that defend territory connected with mate-locating behavior. If this is true—and it appears to be especially so in birds (e.g., the Bower birds)—then Silberglied's observations would explain why bright colors, iridescence, and low variability in color pattern are the rule rather than the exception in male butterflies.

It is also well known now that some butterfly species are sexually dimorphic (have two distinct color forms) for the reflectance of ultraviolet light. Ultraviolet signals are used to recognize conspecific mates in the Orange Sulphur (*Colias eurytheme*) where it is inherited as a (sex-limited) male character on the X chromosome (Silberglied and Taylor 1973). Lepidopterists, of course, are handicapped when it comes to seeing either the long red wavelengths or the short ultraviolet wavelengths that define the limits of the visible light spectrum—we simply do not have the photoreceptors or the brain hardware to interpret these colors.

Once a female is located, a courtship follows that is fairly uniform for all butterflies. For example, the perching males of the Great Copper (*Lycaena xanthoides*) wait for females to enter their visual field. When a male spots a passing female, he darts out and hovers over her, beating the wings in wide sweeps about seven times per second, after which he induces her to land and attempts to mate (Scott and Opler 1975).

The female may reject the male by avoiding his approach, or if induced to land, by flapping her wings in short bursts, while raising her abdomen until the male "ceases and desists." In searching species such as the Checkered White (*Pieris protodice*) a male locates a newly emerged female on her perch. The male initiates a successful courtship by touching her with his wings or legs as he alights on the female. He then approaches the female's thorax in a head-to-head or side-by-side orientation, and curls his abdomen down between her hind wings, grasping her genitalia with the lateral valves (claspers) (Rutowski 1979). Rejection is communicated when the female lifts her abdomen and flutters her wings.

Despite the minor variations, butterfly courtship follows a single basic pattern, with males locating females largely through visual cues (Silberglied 1977). However, researchers have determined that special sexual scents or pheromones—volatile chemical compounds exchanged between male and female—are also required to initiate coupling (e.g., Myers and Brower 1969), probably in most butterfly species. Males may also use auditory and tactile cues (Ross 1963;

Swihart 1967; Kane 1982) in addition to visual cues, and *aphrodisiac pheromones* intended to make the female "settle down" and follow through with appropriate sexual behavior (Grula, personal communication). It is the female that chooses, only the male that can lose at courtship.

Only rarely do females err by accepting a mate from a different species or genus, and very rarely, homosexual couplings or two males coupled to a single female are reported. Mistakes in courtship are sometimes made when the female (usually freshly emerged) is not old enough to discriminate and correctly process the sexual cues of the male. This certainly is the case where dense populations of the Common Sulphur (*Colias philodice*) and the Orange Sulphur (*Colias eurytheme*) overlap, resulting in several different, naturally occurring hybrids (Taylor 1972).

The reliability of visual cues during sexual courtship is complicated in sex-limited mimicry complexes in which only females mimic a distasteful model (Remington 1973), and in species that are polymorphic or exhibit seasonal polyphenism. However, Silberglied (1977) points out that the courtship behavior of all lepidopterans probably involves pheromones to some extent.

In fact, many Müllerian mimics are fragrant to our noses, perhaps indicating not only unpalatability to vertebrate predators, but odors of sexual importance to mates. Pheromones are produced and distributed by a variety of structures, including modified scales and hairs, the so-called androconial patches, and by anal *hair pencils* (in danaid butterflies). Males possessing these scent-producing or distributing structures usually brush them against the female in order to elicit an appropriate response. If visual, pheromonal, and tactile cues are not acceptable for whatever reason, the female either flies away or exhibits the stereotypical rejection posture by fanning the wings and elevating the abdomen.

Coupling or copulation takes place for one-half to two hours, during which time the sperm-containing spermatophore is transferred to the female's bursae. Mating times also vary within and between families of butterflies. Generally, papilionids, pierids, and satyrids mate in the early afternoon, whereas nymphalids and lycaenids have highly variable mating periods from morning until early evening. By contrast, danaids tend to mate in the late afternoon (Miller and Clench 1968; Ferris 1969). However, it should be pointed out that mating periods of butterflies are documented largely by anecdotes. Much more rigorous study is required to understand this aspect of butterfly ecology.

Sometimes mating couples voluntarily, or of necessity, make postnuptial flights. In these cases only one individual flies while the mate hangs on with wings closed. Males typically carry the females in flight in the pierids and danaids, while the female is the active partner in the satyrids. Both males and females are reported to be active flying partners in the Nymphalidae and probably the Lycaenidae as well (Miller and Clench 1968). Postnuptial flights are usually due to disturbances, but the Checkered White makes what appears to be deliberate postnuptial flights, perhaps to move the copulating pair to a less conspicuous and safer location (Brower, Brower, and Cranston 1965; Rutowski 1979).

Females and males of many species may mate more than once, and sometimes this can be determined by dissecting a gravid female and counting the number of spermatophores deposited. The males of *Parnassius* and other butterflies secrete a thick proteinaceous plug—the species-specific sphragis—around the abdominal tip of the female genitalia. This hardening "chastity belt" presumably prevents her from mating more than once, thus ensuring that only the sperm from the contributing male inseminate the eggs.

In one unusual case involving *Heliconius erato* butterflies, males transfer antiaphrodisiac pheromones to females during mating (Gilbert 1976). The odiferous pheromone involved, said to smell like witch hazel, is presented by so-called stink clubs adjacent to the eversible abdominal glands—but only in mated females. The fragrance of this chemical differs between geographic races of *H. erato*, and the hypothesis is that these odors are transferred by the male (where they are contained internally, presumably associated with the sexual apparatus) to the female. The stink-club odor may be contained inside the male clasper, within a pouch lined with glands. Because the odor apparently inhibits additional courtship, the antiaphrodisiac label is warranted. However, this does not explain why the stink clubs are extruded when *Heliconius* butterflies are handled by human captors—perhaps it serves as a defensive smell to more sensitive vertebrate noses than our own.

Monogamy may not be in the best interests of *H. erato* females, because, according to Gilbert (1976), additional spermatophores from later in the female's life probably increase egg production capacity—that is, fecundity. The competitive males of *H. erato*, like several other species of *Heliconius*, "wait" on female pupae about to emerge, sometimes by the dozen. These males are oblivious to other stimuli, apparently concentrating on having the first attempt at mating upon eclosion of the female. When Gilbert brought the abdomen of a mated female to within a few centimeters (about one inch) of waiting-male clusters, the males became agitated and dispersed. Thus, the antiaphrodisiac quality of the stink clubs seems certain, although the odor diminishes with time.

After the spermatophore is transferred to the ostium bursae, it may be absorbed and its nutrient contents incorporated into the female's eggs. Boggs and Gilbert (1979) have documented the male contribution of nutrients to egg production in three species of butterflies: the Monarch (*Danaus plexippus*), *Heliconius hecale*, and *H. erato*, by dosing or feeding males with radioactively labeled sucrose and an algal protein. These males were then mated to virgin females, which in turn were later analyzed along with their eggs.

The first eggs were radioactive in all species, indicating a rapid incorporation of substances contributed by the males during mating. Because unfertilized eggs were about as radioactive as fertilized eggs, the radioactivity of the eggs was not due primarily to fertilization by the sperm. In fact, the radioactive substances were distributed throughout the female's body within a week of mating, perhaps affecting her survival as well as contributing to vitellogenesis. Boggs and Gilbert (1979) suspect nutritional inputs other than the spermatophore as well, and this may explain why female *Heliconius* butterflies mate repeatedly over the length of their extraordinarily long adult life spans.

Many species of butterflies have seasonal forms or "morphs," and in at least one case sexual selection favors these forms differentially, depending on the time of the year (Smith 1975). The species in question, the danaid *Danaus chrysippus*, has at least four color forms in East Africa, two of which, *chrysippus* and *dorippus*, show seasonal difference in mating advantage. The *chrysippus* form has a mating advantage that lasts three to four months, from April to July, but this advantage is lost as their frequency in the population increases. You would expect each form to mate with a frequency that parallels its frequency of occurrence in the population—that is, if no sexual selection is taking place. By July and August, mating advantage appears to shift to the *dorippus* form, and for the remainder of the year neither form has the advantage (Smith 1975).

Since *chrysippus* males are most common during periods of fast population growth, their selection by females may be dependent on their frequency in the population. When they are rare, they have a mating advantage, but when they are abundant, they are selected against. Female *dorippus*, by contrast, almost always have an advantage over *chrysippus* females, whereas male *dorippus* appear to have a mating advantage during the dry periods of July and August, immediately after the frequency of the male *chrysippus* form has peaked. Perhaps then, this case of a *balanced seasonal polymorphism* is maintained in response to differences in mating advantage against a backdrop of seasonal changes. This example helps to illustrate the complexity of the apparently straightforward mating system of butterflies.

Butterfly variation is not limited to sexual differences and seasonal morphs. Aberrations and seasonal and geographic variations in phenotype are commonplace. Ultimately, most aberrations and variations are genetically determined to some degree—directly or indirectly. The environment can modify gene products—proteins, hormones, and enzymes—or regulate their expression. With the development of the techniques of modern molecular biology (genetic engineering), tremendous progress has been made in understanding how the genes of eucaryotic cells—those with a true nucleus and protein-bound chromosomes—are regulated.

Genes are composed of functionally discrete segments of the hereditary molecule, DNA (deoxyribonucleic acid). These genes are located at various positions on the chromosomes—complex nuclear structures composed mainly of DNA and numerous types of proteins that probably serve several structural and regulatory functions. Complex organisms have thousands of genes, all responsible for directing the production of proteins—which can serve as hormones and enzymes, or to make structural components like hair and fingernails. The interaction of certain chromosomal proteins and other DNA elements in the vicinity of genes is probably the most important mechanism for turning the genes on and off at appropriate times in development. There is also evidence for other sophisticated control mechanisms (see Brown 1981 for a review).

However, environmental cues can also affect the regulation of genes, hypothetically turning them on or off in accordance with the type of cue. Indeed, the development and metamorphosis of a butterfly is controlled by sets of genes, which are regulated according to the developmental state of the animal, and environmental stimuli such as photoperiod (duration of daylight), temperature, or humidity. So little is known about gene expression in butterflies that the field is literally virgin territory.

As with nearly all sexual organisms, the genes of butterflies typically come in matched pairs, one on each of two homologous chromosomes. Thus, a gene coding for a specific enzyme involved in the metabolism of melanin, for example, has two forms, or alleles, that both encode the information for the production of the enzyme. The alleles may be identical (homozygous) and thereby produce identical enzymes, or different (heterozygous) and produce slightly or even radically different varieties of the enzyme required by a given chemical reaction in the melanin pathway. While individuals can have only two alleles for each type of gene (one from each parent), populations may have twenty or more alleles of a single gene, greatly increasing the genetic variability of the population.

A sexual organism's body cells are diploid, that is, they contain two of each type of chromosome, each of which contains genes that code for the same set of structures or functions. The diploid chromosome number is halved when the reproductive cells within the ovaries and testes divide to form sperm or eggs. These reproductive cells are haploid because they normally carry only one member of each chromosome pair.

After copulation, the spermatophore is stored in the female's bursae, and the sperm are released to fertilize the eggs passing down the ovarian tubes. Fertilization means not only that one egg and one sperm fuse to form a single cell, the zygote, but also that the chromosomes are matched in the process. Hence, the one-celled zygote, a new butterfly in the making, is diploid—as were its parents.

Immediately, some genes responsible for development of the first instar are "turned on," and subsequent developments in metamorphosis are regulated by some of these early expressed genes. A single gene, particularly one that regulates biochemical reactions, may have a certain effect during one stage, and multiple effects in another. Such *pleiotropic genes* apparently regulate a common step utilized by more than one chemical pathway, hence the multiple effect on the external appearance or phenotype of the butterfly.

But this is hardly the whole story, for we now know that some genes or segments of DNA called transposable elements can "jump around" within a chromosome, between chromosomes, and probably even from one species to another (Lewin 1982). Such elements, especially if they are regulatory in nature, can have dramatic effects on gene expression and phenotype. Butterflies undoubtedly possess such movable DNA sequences.

In addition, eucaryotic genes now are known not to be uninterrupted stretches of DNA, but to consist of regions encoding the information for protein synthesis, called *exons*, and segments which do not contain such information, called intervening sequences, or *introns*. What these introns actually do and how they evolved is the subject of intense study—unfortunately not in butterflies.

After the DNA composing an entire gene is transcribed into RNA (ribonucleic acid), the introns are cut out by specific enzymes that snip the RNA in precise locations. Then, the functional exons are covalently bonded together before the spliced pieces can be expressed or translated into proteins by the chemical machinery of the cell (see Brown 1981 for a review).

The actual situation is far more complicated than this, but my intentions are to point out how little is known about the ways in which genes function. We undoubtedly will be rewriting many textbooks to incorporate the fantastic amount of new information about the genetics of eucaryotic organisms, and I hope that information about butterfly genetics and gene regulation is included.

In the meantime, butterfly genetics is at a very primitive level—their chromosomes are usually very small (Maeki and Remington 1960), and thus they are difficult to observe during cell division and sex cell formation. Virtually no one uses butterflies as models with which to investigate the regulatory nature of DNA. Thus, our present understanding of the genetics of aberrations, sexual, seasonal, and geographic forms, may literally change overnight with the development of new analytical techniques, and the minds to use them. The best we can do now is pigeonhole variation phenomena into categories such as "environmentally induced" and "genetically induced"—categories that not only seem very superficial, but that also may obscure the basic mode of gene regulation underlying a given phenomenon.

Most cases of dimorphism (two morphs) and polymorphism (many morphs), whether seasonal or geographic in nature, are genetically determined (e.g., Hazel 1977). Larvae and pupae also exhibit polymorphism (e.g., Clarke and Sheppard 1972). In fact, a larva's color and pattern often change with each molt, and a population of larvae from a single female may contain several to many distinct color morphs. For example, I have observed that the larvae of the Dainty Sulphur (*Nathalis iole*) vary from a light green to almost a purplish color, with or without longitudinal stripes of different colors.

Another example of color polymorphism is reported for the larvae of the Bordered Patch (*Chlosyne lacinia*) (Neck, Bush, and Drummond 1971; Neck 1976a). Neck and his colleagues have described three color forms: rufa (all orange or orange-red), nigra (all black), and bicolor (basically black but with a row of dorsal orange dots that nearly blend to form a longitudinal stripe). In the bicolor form, the black pigment is in the cuticle while the orange pigment lies below transparent cuticular sections in the deeper hypodermis below.

The color morphs involve two genes on different chromosomes. Each gene, of course, has two alternate expressions or alleles in each individual. For example, a caterpillar that receives a bicolor (B) allele from each parent (thus is genotypically BB) is described phenotypically as bicolor. If the caterpillar receives one (B) allele and one nigra allele (b), which is recessive to (B), then its genotype or genetic makeup is (Bb), and its phenotype is also bicolor. Physically—that is, phenotypically—one cannot tell a (BB) genotype from a (Bb) genotype since both are phenotypically bicolor morphs. If the caterpillar receives both recessive alleles, its genotype is (bb), and its phenotype or morph is always nigra.

The second rufa-controlling gene also controls the expression of the first gene for the expression of the bicolor and nigra morphs, and this makes the inheritance more complicated than indicated above. In the second gene, the rufa allele (R) is dominant to the recessive allele (r) for nonrufa. Hence, (RR) and (Rr) are rufa genotypes whereas (rr) is nonrufa. If (R) is present, however, there is no phenotypic expression of the bicolor-nigra gene. This means that the B and b alleles are suppressed. Thus, (RRBB), (RRBb), (RRbb), and (RrBB), (RrBb), (Rrbb)—all possible combinations of at least one (R) and either (B) or (b) alleles—are all the rufa morph phenotypically. However, a caterpillar with the recessive (rr) genotype allows the first gene

(B) or (b) to be expressed. If it is genotypically (rrBB) or (rrBb), it is a bicolor morph, but if it is completely recessive for both genes (rrbb), it is the nigra morph.

There are some intermediate forms between these morphs, but the nigra, bicolor, and rufa types predominate. In addition, the environment may affect the expression of these two genes—cooler temperatures are known to produce darker, more melanized caterpillars, possibly indicating a thermoregulatory function. Also, crowded conditions may produce darker larvae. This is an exciting system, for it shows very clearly that although there are basically three genetically determined morphs—a discontinuous type of variation controlled by two different genes—that environmental conditions may still modify the phenotype so that variations appear less discrete.

Caterpillars also differ in their biochemistry. Molecular studies of such variation, especially of organic catalysts like enzymes, are useful in determining differences between sibling species—those that look phenotypically alike. Burns (1972) describes such enzymatic variation in *Colias* butterflies. While an individual can have only two alleles for each gene, a population of many individuals may have many different alleles for the same gene, as we have seen. In a single population of Orange Sulphur butterflies, Johnson and Burns (1966) found that one enzyme (dimeric esterase or ESD) is represented by as few as three to as many as twenty-five alleles. But why should a single population of *Colias* butterflies exhibit such a tremendous variation in esterases, each representing different enzyme products?

The Orange Sulphur is not alone in this enzymatic variability at the population level. The Common Sulphur, with twenty-five alleles, and Reakirt's Blue (*Hemiargus isola*) with nine to fifteen alleles also show wide variation in dimeric esterase. First of all, dimeric esterase, as its name implies, is a double molecule (dimer), which means that combinations of the products from different alleles can increase the apparent number of alleles present.

For example, an allele coding for a single enzyme (a monomer) can only produce that specific monomer. Thus, the number of monomers present in a population can only be as great as the number of alleles present. But a dimeric enzyme is produced by combining the products of both alleles—whether they are homozygous or heterozygous. Therefore, an individual with two alleles, say heterozygous, can produce three different enzyme products, but a population with five alleles for the same gene can produce fifteen different dimeric enzymes (or isozymes as they are technically called). Each combination of the products of both alleles is a slightly different enzyme. Thus, an individual with an E_1E_5 genotype can produce the dimeric esterases E_1E_1, E_1E_5, and E_5E_5 and nothing else.

The dimeric esterase system is easy to study because males leave a spermatophore in the female after mating, instead of a liquid ejaculate. Thus, larvae can be reared from wild-caught females found later to have mated only once. Butterflies can also be anesthetized by carbon dioxide and hand paired for more exacting control of the genetic studies. Furthermore, ESD studies are facilitated because the isozymes are found in all developmental stages from egg to adult.

The results of Burns's investigations show that most *Colias* species have multiple alleles for the ESD gene. But this was hard to explain by some population geneticists who used mathematically nonadditive models in an additive way to predict that genetic variability should be kept to a minimum. So, their predictions resulted from the assumption that each gene was selected individually and independently of other genes. Actually, the entire butterfly—the whole gene package—is the unit upon which natural selection normally acts, although some very well known exceptions exist. It is possible that the large amount of variability at the esterase locus (the locus is a gene's position in the chromosome) is of no selective value—it is selectively neutral. Others claim that high variability is an anachronism—a relict of past selective events, or that high mutation rates at the ESD locus are responsible for the great variability (Burns 1972).

No one really knows what events maintain such high ESD variability in present-day *Colias* populations. Population size, dispersal patterns, and mating systems are all important in maintaining existing variability. Perhaps the gene for esterase is easily mutated and perhaps the esterase variations have little selective value, good or bad, but a more likely possibility is that esterase variability is important to *Colias* populations inhabiting heterogeneous environments—environments that vary with, say, the chemistry of the larval food plant.

Burns, Johnson, and others have used *Colias* species to investigate the importance of protein variability, how protein variability is maintained, and to measure population differentiation. *Colias* was chosen because it had the necessary criteria for experimentation—it is an intensely studied, common group of what appear to be morphologically conservative species, whose species delineations are often indistinct and probably frequently crossed.

Yet, *Colias* is a "difficult" genus because not only do small populations of highly variable individuals exist, including hybrid swarms, but there is commonly extensive variation seasonally and between the sexes. In short, it is the perfect enigmatic group that scientists love to study. Burns (1972) has pointed out that North American *Colias* species can be divided into three distinct groups based roughly on food plant as well as on esterase activity. These include the following: the *Vaccinium* eaters (low ESD variation); the *Salix* eaters (moderate levels of ESD variation), and the legume eaters (with very high levels of ESD variation).

Burns shows that the pattern (quantity and type of enzyme) of ESD variation is sufficient to identify an individual at the species level. Queen Alexandra's Sulphur (*Colias alexandra*) and *C. harfordii* all had high levels of variability, with single populations having fourteen to seventeen different alleles, but the Orange Sulphur had at least twenty-five, possibly thirty different ESD alleles! In contrast, the *Vaccinium* eaters such as Behr's Sulphur (*C. behrii*) and the Pink-edged Sulphur (*C. interior*) had only three and four alleles respectively. Why should legume eaters have a tremendous esterase variability and *Vaccinium* eaters very little?

Obviously, some sort of selection is operating here, possibly because the glacial relict species (*C. behrii* and *C. interior*) are Nearctic species confined to restricted habitats whereas *C. eurytheme* is not only common and widespread, but continues to expand its range. Thus, perhaps species like *C. behrii* show little variability because they have small populations, restricted habitats, and limited food plant choices, whereas *C. eurytheme* shows great variability because it occupies a broad and highly variable geographic range and feeds on many different food plants within that range.

Pupae may also exhibit polymorphic traits. For example, the pupae of pierids and papilionids are often characterized by striking color polymorphisms. Wiklund (1972) and many others have investigated pupal polymorphism in the Papilionidae. In Europe, the Old World Swallowtail (*Papilio machaon*) is a favorite test species (e.g., Wiklund 1975a), whereas in the United States, the Black Swallowtail (*Papilio polyxenes asterius*), in the same general taxonomic subgroup as the Old World Swallowtail, is used (e.g., West, Snellings, and Herbeck 1972). The pupal colors of the Black Swallowtail vary from green and yellow, to brown and white, to a nearly black form. Wiklund (1972) showed that nearly 100 percent of green and yellow pupae could be produced when last instar larvae were exposed to sources consisting of largely one wavelength (monochromatic light), including wavelengths of 550 nanometers (green), 580 nanometers (yellow), and 620 nanometers (orange).

However, only 55 percent of larvae formed green-yellow pupae in complete darkness, the remainder being brown-white—a value not significantly different from the percentages of green-yellow to brown-white pupae produced when larvae were exposed to 460 nanometers (blue), 510 nanometers (blue-green), 670 nanometers (red), 750 nanometers (infrared), and 850 nanometers (far infrared). Clearly, larvae are sensitive to wavelengths of light incident upon them prior to pupation, and this effect is often viewed as a color adaptation to the background color of the pupal site.

The story of pupal dimorphism and polymorphism is slightly different with the Cabbage White (*Pieris rapae*), as discovered by Bernath (1981). The pupae of the Cabbage White vary from a clear light green through an intermediate green-brown to a dark gray-brown. This variation is continuous but commonly referred to as a polymorphism, since humans can discriminate different morphs with some degree of repeatability. The pupal color has two basic components: a ground color that is basically green or brown as in the Black Swallowtail and the Old World Swallowtail, and a overlying melanization that modifies the appearance of the ground color.

The ground color insectoverdin (literally, "insect green") results from a combination of yellow lutein in the cuticle and mesobiliverdin in the epidermis, while the brown color is due to xanthommatin in the epidermis. Melanin occurs in the cuticle of both green and brown forms. Bernath divides the degree of melanization into major dots and minor dots—those that are visible by eye, and those too small for human resolution. Major dots occur along the dorsum and seem to correspond to sites of muscle attachment in the adults.

Minor dots appear as a faint dusting and serve to darken the ground color.

By using a technique known as photoacoustic spectroscopy—first invented over one hundred years ago by Alexander Graham Bell—Bernath measured the spectra of living pupae. She found that the variation in ground color was energetically unimportant in the absorption of sunlight. However, the amount of energy absorbed doubles or triples as melanin deposition increases, and possibly this might be of some minor thermoregulatory importance. Bernath set out to find a consistent set of environmental cues that might be responsible for the pupal color of the Cabbage White. She examined the effects of photoperiod, temperature, substrate transparency, and month of pupation on first and second instars.

All treatments fell fairly neatly into two categories: (1) less than 10 percent brown pupae produced per treatment and (2) 30 to 70 percent brown pupae produced. Green light is a green-inducing pigment stimulus whereas blue and ultraviolet are brown inducing. The wavelengths responsible for brown pupae were between 350 and 450 nanometers. In addition, blue paper substrates are brown inducing while both green and brown papers are green inducing. These results are explained by spectroscopy—brown paper has the identical amount of absorbance as green, and brown-inducing cues are basically a suite of cues associated with open sites (e.g., away from green vegetation), dark sites, or with fall conditions—when pupation sites are or will be predominantly brown.

Bernath found that the best environmental predictors of pupal color are, in order of decreasing importance: month, photoperiod, temperature, substrate color, and substrate transparency. For example, the pupae of summer are usually green and the pupae of autumn are usually brown. Apparently, there is no thermoregulatory advantage for brown pupae. However, for heavily melanized pupae that overwinter, there may be some, although minor, heating advantage in the spring, especially since the Cabbage White is among the earliest butterflies to appear, along with hibernating species such as the Mourning Cloak (Nymphalis antiopa).

Proof for this conjecture awaits testing. Bernath (1982) discovered, however, that color can be important for crypsis in some cases (for example, green pupae on brown sticks). Cryptic pupae, whether green or brown, had a longer survival time than noncryptic pupae. Furthermore, there is a seasonal parental effect so that more brown pupae are produced in the autumn. Selection, then, is always heaviest against pupae that were mismatched for the color of their background, which explains at least in part the pupal polymorphism in the Cabbage White (see also Clarke and Sheppard 1972).

Because adults are flashier than larvae and pupae, their polymorphisms and seasonal polyphenisms have been studied more extensively. For example, Aiello and Silberglied (1978b) determined that the orange and red hind wing markings of the Fatima butterfly (*Anartia fatima*) are inherited in a simple Mendelian way. Butterflies with aberrational orange genotype are designated (rr), while the normal or wild type genotype with red hind wing markings are either (Rr) or (RR). Thus, the red allele (R) is dominant over the orange allele (r). The gene for red or orange markings in *Anartia fatima* is carried on one of the "nonsex" or autosomal chromosomes (autosomes). The autosomal chromosomes of males and females are identical, matching pairs, and hence both males and females have equal chances of inheriting orange or red markings on their hind wings.

Sometimes variations are claimed by indirect evidence to have a genetic basis when in fact they do not. One such case involves the color variation in the bands on the dorsal surface of *A. fatima*, which appears to be phenotypically yellow or white. At one time it was thought that yellow-banded females had a mating disadvantage and that the yellow and white individuals represented a balanced polymorphism that is maintained in both sexes. However, Taylor (1973a) hypothesized that only one phenotype actually existed—a yellow one which faded within several days to white from exposure to sunlight. Thus, the older individuals tended to be white while newly emerged individuals tended to be yellow.

Females with yellow wings were typically newly emerged with large fat bodies and lacked mature eggs. Females with white wings had depleted fat bodies, worn wings, but were producing mature eggs. And interestingly, age changes the ultraviolet reflectance pattern on these bands as well. Yellow bands were ultraviolet absorbing while cream and white bands were ultraviolet reflecting. However, there are apparently no distinct ultraviolet phenotypes—they formed a continuum from ultraviolet absorbing to ultraviolet reflecting.

Anartia fatima males approach other males even more often than females, which is puzzling, since it seems that males should chase yellow-banded females with the most reproductive potential. It is even possible that some genetic factor does have some bearing on the color of the wing band, but it is not possible to accurately determine those from individuals who have faded from yellow to white (Taylor 1973a).

Variations that are rare and unusual are generally not called forms or morphs, but *aberrations,* even if these variations are genetically determined. Actually, if one desired, every butterfly could be an aberration of some hypothetically pure form, since no two butterflies are exactly alike. Variation, in fact, is universal among all living organisms—a hallmark of life. Unfortunately, some lepidopterists go off the deep end when it comes to naming species, subspecies, forms, aberrations, and other sorts of phenotypic variations that are easily discernible to the human eye. Many of these names are scientifically useless and invalid. Worse, they make one lose sight of the universality and evolutionary importance of variation. The time spent fractioning forms could be better spent investigating the reasons for the variations.

One common type of aberration is the appearance of heavily melanized specimens, often with very different markings than the normal phenotypes. There are literally hundreds of these reported aberrations (e.g., Davies and Arnaud 1967). A possible explanation is that developmental mistakes occur at the genetic level in the epidermal cells producing wing pigments, or in the focal areas responsible for wing pattern. What precisely causes such aberrations, however, is for the most part unknown.

Another remarkable aberration is the mutant of *Danaus plexippus erippus,* the South American counterpart to the North American Monarch butterfly. Clarke and Rothschild (1980) report that this mutant has a wing cell on both ventral and dorsal forewings that is yellowish-cream color instead of orange. Their breeding experiments indicate that this aberration is due to an autosomal recessive gene: a recessive allele located on an autosome. Thus, it can be expressed in both sexes as long as two recessive alleles for the trait are inherited. Matings between these mutants are nearly infertile, and it is likely that the mutation in this case affects more than wing color (e.g., it is pleiotropic and affects the fitness of those bearing the alleles). Like most mutations, this one is apparently not advantageous to survival.

There are some spectacular aberrations in which a butterfly may possess both male and female characteristics. These gynandromorphs are sometimes bilaterally symmetrical so that one half of the body—including the genitalia—is female, while the other half of the body is male. *Bilateral gynandromorphs* are rare and highly prized by collectors. A gynandromorph of the Diana Fritillary (*Speyeria diana*) was reared by Showalter and Drees (1980) that is unusual in several ways. Recall that males of the Diana Fritillary emerge about two to four weeks before females because they are smaller and develop faster. Females reportedly fly in concert with their distasteful Batesian model, the Pipevine Swallowtail (*Battus philenor*), and this fact may explain why females emerge later than males.

The researchers discovered that the rate of development of the gynandromorph caterpillar was intermediate between that of males and females as groups, possibly because the larval developmental rate was half male, half female. The chrysalid also showed external features characteristic of both females and males. The right eye (male) of the adult is larger than the left eye (female), yet the right wings are unusually large for males, while the left wings are unusually small for females (Showalter and Drees 1980).

The sexual phenotype of each cell in a butterfly's body is determined directly by its genetic constitution, whereas with mammals such as ourselves, the genetic constitution of the fertilized egg determines the sex of the reproductive organs—testes for males, ovaries for females. The secondary sexual characteristics typical of mature adults are determined by an interplay of powerful hormones secreted from specific areas in the brain and within the reproductive complexes. These hormones are distributed via the blood, thereby affecting the total phenotype, making it either male or female.

Hormonal imbalances in humans can lead to masculine-looking women and feminine-looking men. But gynandromorphic butterflies probably most commonly result from the fertilization of an egg that has two nuclei, hence two haploid sets of chromosomes, one of which contains an X chromosome, the other of which contains a Y chromosome. Since a butterfly male is designated as (XX), it can only donate an X chromosome, whereas females (genetically

[XY]) can contribute either X or Y. Normal zygotes are either (XX) or (XY), but if a single cell with two nuclei (one [X] and the other [Y]) is fertilized by sperm containing only X chromosomes, then its genetic constitution becomes both male (XX) and female (XY). The side that develops from the (XX) genotype is male, while the side that develops from the (XY) genotype is female.

Gynandromorphs can also be produced from an (XX) zygote that divides into two cells, one of which loses an X chromosome. This leaves a two-celled embryo with one (XX) cell and one (X__) cell. Since it is the number of X chromosomes that seems to determine sex, not the presence of the Y chromosome, the (XX) cell develops into a normal male side, while the (X__) cell becomes female more by default than anything else. Because this type of cell division accident may occur at any stage of metamorphosis, some butterflies may be gynandromorphic in one wing only, thus producing a "sexual mosaic."

Outside of sexual dimorphism, phenotypic polymorphism, and variations due to peculiarities in gene expression as well as genetic accidents during cell division involving sex chromosomes, seasonal polyphenism is responsible for much of the variation within a butterfly species. A polyphenism strictly interpreted is variation that is of nongenetic origin within a population. A seasonal polyphenism is a seasonal change in phenotype, usually fairly discrete rather than continuous, within a population. This means that if larvae from a single female are divided into two groups and then exposed to different but appropriate environmental stimuli, they will develop different phenotypes even though only one genotype is present in the entire brood (e.g., Shapiro 1973; Douglas and Grula 1978).

However, we should take nongenetic variation with a grain of salt, since we do not understand how gene function and expression is regulated. Perhaps different environmental conditions activate or suppress the products or expression of some genes—implying that different genes or their differential expression are involved in seasonal polyphenisms. Still, the different phenotypes are expressions of the same genome in different environments.

Seasonal polyphenic forms can be so strikingly different that even experts have been fooled into believing that each form represented a distinct species. Perhaps the most famous European example of seasonal polyphenism is that of the nymphalid butterfly *Arashnia levana,* which has a fritillarylike spring or *vernal morph,* and a strikingly different summer or *aestival morph.* The seasonal polyphenism is determined primarily by changes in temperature experienced by the last instar larvae and pupae.

There are many other examples of seasonal polyphenism, some spectacular, others less so, but each illustrates the effect of the environment on phenotype, particularly environmental stimuli such as the intensity and duration of sunlight, availability of water, and fluctuations in temperature. For example, the Pearl Crescent (*Phyciodes tharos*) has a seasonal polyphenism of a discontinuous variety, in which larvae exposed to long days give rise to summer adults (form 'morpheus') with less intensely colored wings than those exposed to the short days of spring and fall (form 'marcia') (Oliver 1976). The adaptive significance of the polyphenism is unknown, but crypsis tuned to seasonally changing backgrounds may be important. The darkening is not likely to be of thermoregulatory importance because the increase in melanization is largely peripheral, not basal.

A similar seasonal polyphenism occurs in the Goatweed Butterfly (*Anaea andria*), in which dark winter forms are produced by exposing fifth instar larvae to short photoperiods and lighter summer forms are produced when larvae are exposed to long photoperiods (Riley 1980). The increase in melanization may be of thermoregulatory significance, but the polyphenism also entails longer-tailed forms with forewings that are more deeply curved and pointed. The Question Mark butterfly (*Polygonia interrogationis*) and the Comma butterfly (*P. comma*) also experience a similar seasonal change in wing coloration and angular outline of wings and tails.

Some seasonal polyphenisms are not induced by changing photoperiod or temperature, but by changes in rainfall. The summer form, 'sara,' of the Sara Orange Tip (*Anthocaris sara*) is produced from a small proportion of the offspring of female 'reakirtii' that, unlike overwintering chrysalids, spend less than three weeks in pupation.

In this case, both rainfall and possibly sunlight may be involved. Even more interesting is the fact that both 'sara' and 'reakirtii' can be produced from the same batch of eggs (Evans 1975), and that in the drier, western part of its range, another form, 'rosa,' occurs in the late fall. Thus, the seasonal forms of *A. sara* may be dry and wet season responses, similar to the dry and wet season forms of the Dog-face butterfly (*Colias cesonia therapsis*), a common Neotropical butterfly whose range extends into the southern United States (Masters 1969).

Often, pupal diapause and adult phenotype are tightly linked. For example, in the Mustard White (*Pieris napi*), diapausing pupae give rise to vernal phenotypes while aestival phenotypes emerge from nondiapausing pupae. However, Shapiro (1977) discovered that pupal diapause and adult phenotype could be decoupled in *P. n. venosa.* Thus, populations apparently contain pupae that must diapause (obligate diapausers) as well as pupae that may or may not diapause, depending on their response to immediate environmental conditions (facultative diapausers).

Facultative diapausers typically respond to day length. Sometimes photoperiods that should induce diapause can be overridden by exposure to high temperatures during development. But larvae exposed to long-day conditions will not diapause even if exposed to low temperatures. However, if long-day pupae of *Pieris napi venosa* are chilled, a vernal phenotype is produced (Shapiro 1977, 1978a).

Shapiro suggests that weedy colonizers—those that must disperse to find new habitat, usually disturbed—can override the normal photoperiodic determination of diapause and squeeze in one more brood if fall weather permits (e.g., Shapiro 1978b). However, vernal phenotypes are the typical product of diapausing pupae, and from nondiapausing pupae exposed to even a few days of cold weather. The same is true of Rocky Mountain populations of the Clouded Sulphur (*Colias philodice eriphyle*), and the darkened, more melanic wings of the vernal phenotype may be useful for thermoregulation under cool conditions.

Sims (1980) discovered that there is a latitudinal gradient in the photoperiodic response in the California race of the Anise Swallowtail (*Papilio zelicaon*), which probably colonized the state in the eighteenth century when it shifted host plants. In temperate and high-latitude pierines, facultative diapause seems to be under photoperiodic control, whereas the influence of temperature affects inhibition (Shapiro 1980b, 1980c). By manipulating temperature and photoperiod exposure to pupae of the *Pieris napi* group and the *P.*

callidice-occidentalis group, Shapiro (1976a, 1976b) may have unraveled the historical range and evolution of these butterflies (Shapiro 1980a).

Shapiro (1971) showed that there was a latent polyphenism in the West Virginia White (*Pieris virginiensis*) that was indistinguishable from that of its multivoltine relative *P. napi oleracea*. By contrast, the Mourning Cloak (*Nymphalis antiopa*) has an enormous geographic range (but an extraordinarily stable phenotype) and fails to produce the aberrant phenotype *hygiaea* when subjected to cold-shock treatment that induces this phenotype in lowland California Mourning Cloaks (Shapiro 1981c).

Shapiro (1975c, 1977) infers from his experiments that in the Nearctic Pierini, univoltinism and monophenism were secondarily derived from multivoltine, polyphenic ancestors. In the Andean genera of Pierini (about forty species), Field and Herrera (1977) have found evolutionarily plastic species whose boundaries are poorly defined, as well as complex zones of integration and polymorphisms. Shapiro (1980b, 1980c) contends that despite the close relationships of some Andean species their polyphenisms appear to have evolved independently. Further field and laboratory work is being done in this area of great interest. It is clear, however, that there is redundancy and variability in the phenotypic induction mechanisms of at least some Pierines (Shapiro 1978a, 1982).

Another excellent example of seasonal pigment polyphenism in sulphur butterflies was determined to be of thermoregulatory significance by Watt (1968, 1969). Many species and isolated populations of *Colias* butterflies occupying cold or seasonally cold climates have heavily melanized hind wings. Watt showed that these melanic morphs absorb sunlight more efficiently than nonmelanic morphs, thereby elevating thoracic temperatures more quickly and to higher levels under cool environmental conditions.

Furthermore, Watt showed that sulphur butterflies bask laterally at thoracic temperatures of 34 to 35°C (93 to 95°F). Watt elegantly illustrated that in alpine or northern areas, melanic phenotypes of *Colias* likely have greater reproductive success than their lighter-colored phenotypes since high body temperatures translate into greater flight and foraging efficiency, opportunity for finding mates, and increased fecundity (Watt 1968).

Variation between isolated subspecies of butterflies is probably not controlled so much by present ecology as by historical factors that contributed to the present-day geographic isolation. For example, Bowden (1979) points out that there is no evidence at present to show that the variations between American and European populations of *Pieris napi* are adaptive, since equal opportunity for crypsis or thermoregulation are possible with either type of veining. However, melanic markings over veins probably have little if anything to do with thermoregulation (see chap. 4). In *Pieris rapae*, normally a dorsal basker, only basal areas on the upper wing surfaces can aid in thermoregulation. Thus, the variation in vein darkening on the ventral surfaces probably has no effect on the ability to elevate body temperature.

Although variation occurs in all butterfly species, it has been most thoroughly examined in the Pieridae. Some of this variation is due to *interspecific* crosses—uncommon, but not unusual. Many hybrids are at least partially fertile when crossed back to either parent species, but there is a great deal of variation, perhaps because the regulation of the genes within the hybrid is upset by two slightly different sets of chromosomes, especially the genes controlling development and metamorphosis.

Oliver (1979) feels that much of the driving force of speciation may in fact be due to small differences in metabolic integration. If incompatibility occurs at the metabolic level, speciation—the splitting of one species into two or more distinct species—may take place with very little genetic differentiation. This may be true for the European White (*Pieris callidice*) and the Alaskan Western White (*P. occidentalis nelsoni*), which are phenotypically similar to each other but physiologically and genetically very different. Other phenotypically similar butterflies, including Central Asian *P. c. amaryllis*, may not have diverged recently from a common ancestor but rather evolved a similar wing pattern in parallel with those of *P. o. nelsoni* and *P. callidice* (Shapiro 1980b, 1980c).

Variation is such a fundamental characteristic of life that we tend to pigeonhole it at different levels. Thus we can categorize sexual variation, geographic variation, and seasonal variation, for example. At some level—and this level differs between different schools of thought—scientists group similar organisms into evolutionary units—the species. A single species may exhibit a great deal of phenotypic variation between sexes, races, and seasonal forms, but if all variations can be crossed and produce viable and fertile offspring, it is nonetheless a single "good" species by the *biological species definition* (Mayr 1963).

Thus, the species is said to be a cohesive and common gene pool held together by gene flow between breeding individuals. But Ehrlich and Raven (1969) have suggested that gene flow is actually very restricted even among populations we consider on the basis of similar phenotypes to represent a single species. Is the species then a natural evolutionary unit held together by gene flow? Perhaps natural selection is both a cohesive force and a disruptive force in evolution, and the evolutionary units are really the locally breeding populations whose size is variable. Such populations (including subspecies) may differentiate under different selective forces quite rapidly.

The dynamics of speciation can be observed in the *Colias* butterfly complex, *C. philodice* and *C. eurytheme* and their hybrid forms. Although these two species currently occupy roughly the same ranges, they were apparently disjunct or geographically isolated over much of the continent prior to the arrival of European agricultural practices (Ae 1959; Taylor 1972). *C. philodice* formerly occupied the North and East, and subspecies were found in the mountainous regions of the West, while the typically orange *C. eurytheme* occupied the southern plains and the Southwest. The area of species overlap, or *sympatry,* has increased steadily so that even in Arizona, where the cultivation of alfalfa brought the two species together, *C. philodice* has become established in nearly all *Colias* populations, and a significant number of interspecific matings occur.

When two previously isolated but closely related species come into contact, hybridization often results. It might be predicted that selection would reduce hybridization because the progeny from interspecific crosses usually have reduced viability and/or fertility, and therefore are less "fit." As a result, the two species should evolve better premating reproductive isolating mechanisms, and eventually interspecific matings would no longer occur. At that point they would be considered two "good biological species" according to the definition of Mayr (1963).

However, if two previously isolated species have not diverged to the point of genetic incompatibility, there should be increasing frequencies of hybridization (Taylor 1972). Taylor found that wherever

C. philodice and *C. eurytheme* overlapped, mating was not random, but occurred with increasing incidence of interspecific mating as density of available females increased. In dense populations, virgin *C. eurytheme* females less than sixty minutes old may be approached by *C. philodice* males. Mating may take place because young *C. eurytheme* females have not acquired the behavioral repertoire and/or perceptual ability to discriminate *philodice* males from *eurytheme* males. However, hybrid frequencies are usually less than 10 percent, and neither species has apparently lost its "distinctness" even though hybrids from the above cross (which is the most common) mate preferentially with *philodice* and other hybrids. In effect, genes are flowing largely one way in sympatric populations, that is, introgressing from *C. eurytheme* into *C. philodice* (Grula and Taylor 1980b).

This is really quite amazing because males of both species are present when females emerge. What mechanism is maintaining the distinctness of these two closely related *Colias* species? Taylor later determined that mature females with their eyes painted black, mated selectively with males of their own species. This indicates that mate selection is accurate even with visual cues absent. Removing the last two-thirds of the female's antennae (a radical antendectomy?) reduced mating by *C. eurytheme* females after thirty minutes, but increased mating by *C. philodice* females.

Thus, the primary defect in the *Colias* mating system that permits hybridization is that newly emerged females cannot discriminate *conspecific* males (males of their own species). Interspecific matings between *philodice* males and *eurytheme* females produce hybrids with reduced fertility and viability. It would be expected that reproductive isolation will improve between the two species, because selection should favor females that most efficiently select conspecific mates. This could be done by delaying mating until females can unambiguously discriminate between the males (Taylor 1973b).

Is olfaction the only cue used to discriminate mates in this mating system? Silberglied and Taylor (1973) determined that ultraviolet reflectance in male *eurytheme* wings and the lack of it in males of *philodice* may help isolate the species during courtship. Instead of the wide range of pigment-produced (visible wavelength) color variations found in both species of *Colias* (including the white 'alba' females—a morph whose frequency varies between popula-tions and between seasonal broods), there are only two phenotypes under ultraviolet light: ultraviolet reflecting (UVR) and ultraviolet absorbing (UVA). All *eurytheme* males from all described popula-tions are UVR, even yellow-colored forms from Mexico, whereas all pure populations of *philodice* are UVA. Alba females reflect ultra-violet slightly, but are never iridescent like *eurytheme* males.

The two researchers discovered that ultraviolet reflection is in-herited as a sex-linked recessive, meaning that it is controlled by a recessive allele carried on the X chromosome. Thus, males of *eu-rytheme* are homozygous recessive $(X_u X_u)$ whereas males of *phi-lodice* are homozygous dominant $(X_U X_U)$. In Arizona, where Taylor conducted his first studies and where interspecific mating is fre-quent, sympatric populations contain yellow (*philodice*-like) but ul-traviolet reflecting individuals, and orange (*eurytheme*-like) but ul-traviolet absorbing individuals. This suggests that the alleles for ultraviolet reflectance and the alleles for normal pigment color seg-regate independently during sex cell formation and are therefore on different chromosomes.

Later, Silberglied and Taylor (1978) studied the courtship of both species under field conditions. The courtship appears very brief and simple compared to the courtship behavior of most butterflies. Males seemed to rely on visual cues to locate and identify females, even attempting to mate with colored paper dummies.

In *eurytheme* males, ultraviolet absorbance of the females inten-sified the approach while ultraviolet reflectance from competing males inhibited any approach. The female *eurytheme* allows a male to approach and mate when ultraviolet reflecting signals are re-ceived from the wings of male *eurytheme*. However, the visible (pigment) colors play no role in courtship or mate selection in either species. *Colias philodice* females do not respond positively to any ultraviolet reflecting patterns.

When hybridization occurs, all first-generation male offspring (F_1) are UVA. In addition, all F_1 hybrid females from a female *C. eu-rytheme* and a male *C. philodice* prefer to mate with males of *C. philodice*, thus reinforcing the introgression of *eurytheme* genes into the *philodice* lineage. However, F_1 females from a *eurytheme* male by *philodice* female cross prefer to mate with *eurytheme* males. Thus, introgression may be asymmetrical because the most common interspecific is a *philodice* male–*eurytheme* female cross (Silberglied and Taylor 1978; Grula, personal communication).

Females of both species do rely on chemical cues for conspecific mate recognition as well, but it was not known to what extent until Grula (1978) continued the investigation into the nature and inheritance of male pheromones. The most important *C. philodice* wing compounds were three different n-hexyl esters, controlled by one or more autosomal genes that are probably dominant. By contrast, the X chromosome carries most of the genes controlling production of the most important *C. eurytheme* compound, 13-methyl heptacosane. In addition, behavioral studies have shown that the X chromosome is the location of most of the genes controlling female response to male courtship signals in both species (Grula and Taylor 1980b). Thus, it seems that nearly all the genes responsible for the inheritance of the courtship communication system are inherited as a complex on the X chromosome in *C. eurytheme*, but are found on both autosomes and the X chromosome in *C. philodice* (Grula and Taylor 1980c). Females of *philodice* respond positively only to olfactory cues—pheromones—while those of *eurytheme* must have visual cues—ultraviolet reflectance—as well as olfactory cues (Grula and Taylor 1979).

Further investigations by Grula, McChesney, and Taylor (1980) and Rutowski (1980) showed that the pheromones are located on the wings and suggested that a species-specific blend of pheromonal components is necessary in courtship. Yet, the pheromones of *C. eurytheme* and *C. philodice* are qualitatively very different, unlike the close chemical similarity found in the pheromone composition of other lepidopterans that are closely related. More sensitive bioassays will be necessary to clarify the extent to which each pheromone is capable of eliciting an appropriate behavioral response.

Equally interesting, these compounds are the largest yet discovered with a pheromonal function in the Lepidoptera. All three esters are liquids at room temperature and can be perceived by humans at close range. The other long chain compounds are solids or liquids at room temperature, which seems unusual, since they must somehow diffuse through the air or make contact with females directly in order to be effective. Perhaps high volatility is not a requirement (or perhaps higher temperatures of the field are sufficient to make them volatile), since the pheromones are used at extremely close range to induce the female to settle and display receptive behavior (Grula, McChesney, and Taylor 1980).

It is at least intuitively satisfying that there is no such thing as an invariant characteristic at any level of organization, from genes to pheromones, isozymes to color patterns. Variation is the raw material of evolution, and although the gross similarities that we perceive between individuals within breeding populations are useful in classification, they sometimes tend to obscure the central themes of life: survival, reproduction, and speciation. The only aspect of life that seems invariant is the capacity for continual change.

CHAPTER 8

The Coevolution of Butterflies and Their Host Plants

Butterflies have been intimately associated with their larval and adult host plants for tens of millions of years (Ehrlich and Raven 1965). As larvae they have influenced plant characteristics such as chemical composition and leaf shape. As adults, flower-visiting species extract nectar and in some cases pollen, thereby influencing flower shape, phenology, and even the reproductive strategies of plants. While many species of butterflies are effective pollinators of flowering plants, they do not necessarily pollinate the flowers of their larval food plants. What we have then, is a coevolving system on many fronts that includes not only the predators and parasites of butterflies, but the larval and adult host plants as well.

The term *coevolution* was conceptualized and popularized by Ehrlich and Raven (1965). Coevolution (formerly termed *coadaptation* in the broad sense of the term) occurs when populations of different species (e.g., plants and their insect pollinators) interact so closely that each population produces a strong selective force on the other. Over time, mutual adaptations in both populations make the relationship more complex. For example, giraffes and their acacia food plants are a coevolving system; so are humans and the two species of symbiotic mites that inhabit the sebaceous glands and hair follicles of our forehead region (Camin, personal communication).

With butterflies, coevolution with other groups is the rule rather than the exception since butterflies occupy an important intermediate level between their larval and adult food plants and their predators and parasites. Thus, on a grander scale, butterflies are coevolving with a community of local organisms. In some cases, the presence of butterflies can adversely affect the existence of other organisms. For example, Blakley and Dingle (1978) report that the activity of the Monarch butterfly (*Danaus plexippus*) larvae has eliminated the milkweed bug *Oncopeltus* from a Caribbean island. This may seem like an insignificant microevolutionary event, but it is the sum total of similar events that may in time alter the community structure and provide the pathway for coevolution.

In the broadest sense, the entire biosphere has been evolving for nearly 3.5 billion years. But the coevolutionary actions considered here are only those very intimate interactions and coadaptations that have developed between butterflies and their host plants. Other coevolutionary adaptations such as between Batesian and Müllerian mimicry complexes, and between lycaenid butterflies and their attendant ants have been discussed in chapter 6.

As interesting as coevolutionary associations are, the interactions of plants and their associated butterflies have only recently received significant attention. Yet, it is not possible to understand the complex ecology of butterflies without understanding these interactions between butterflies and their host plants. Until recently, then, we have observed butterflies as though they evolved in a vacuum.

Butterflies are often said to be coevolving with their larval host plants (e.g., Feeny 1975) as well as with their adult host plants, and certainly this must be so. But the extent and degree of this coevolution is obscure and poorly documented for the most part. We list new host plants for species of butterflies as if there were some sort of race, but rarely do we compare the extent of use, or the success of larvae reared on these plants.

We have, at least, tried to categorize butterflies according to the range of the host plants that they utilize. Species that utilize only one species of larval food plant in a given geographic area are said to be *monophagous* (e.g., see Levins and MacArthur 1969). Others are *oligophagous* and use several different species within a family of plants, while polyphagous species may develop normally on dozens of species from perhaps a dozen different families of plants (Slansky 1974). However, definitions regarding these three terms may differ considerably among specialists (Bower, personal communication).

Although definitions may vary, larval host plant preferences or ranges should not be used as a basis for separating butterfly species. For example, the larval hosts of the Dorcas Copper (*Lycaena dorcas*) and the Purplish Copper (*L. helloides*) include the genera *Rumex*, *Polygonum* (Polygonaceae), and *Potentilla* (Rosaceae). The Polygonaceae and Rosaceae are taxonomically distinct as far as plant taxonomists are concerned, but the three genera within these two families share a specific flavonoid compound that apparently acts as a chemical signal to induce oviposition by females of both lycaenid butterflies. Thus, although both species are essentially oligophagous on Polygonaceae, the independent occurrence of a common flavonoid in both plant families has allowed the butterflies to diversify their larval food sources and become polyphagous in the process.

In an earlier study, Downey and Fuller (1961) concluded that the Common Blue (*Plebejus icarioides*) is an oligophagous lycaenid that utilizes sixteen species and seven different varieties of lupines (*Lupinus*). However, the Common Blue might also be classified as monophagous because in a given geographic area only one species of *Lupinus* is utilized. Of course, it is the adult butterfly that chooses the food plant for the larvae, and perhaps the spotty and localized distribution of the butterfly is partly due to the spotty and localized distribution of *Lupinus* species. This may be responsible for the great range of phenotypic variation that exists between isolated colonies of this butterfly.

Nutritionally, though, many species of *Lupinus* are satisfactory for rearing larvae, not just the *Lupinus* species chosen in a specific area. As a group the lycaenids may be more catholic in their choice of food plants because many species are flower feeders, whereas most butterflies outside the Lycaenidae are leaf feeders. Still other lycaenids consume flowers in the early instars and migrate to young leaves in later instars. Perhaps flowers have fewer plant toxins and are chemically less distinct and threatening than are the leaves of many plant species.

The Common Blue prefers to oviposit on pubescent species or varieties, so that whenever two or more potential *Lupinus* hosts are available, the hairiest phenotype is preferred, possibly identified by tactile receptors on the foretarsi rather than by chemical cues. Nutritional value apparently has no bearing on the choice of larval host plant.

In one case where two *Lupinus* species are utilized, associated ants are believed to transport the caterpillars from one species of *Lupinus* to the other. Adults reared on one *Lupinus* species will not necessarily choose that species for oviposition, especially if more pilose species are present. The restriction to species of *Lupinus*, however, probably has a genetic basis (Downey and Dunn 1964). Whatever the reason for differential selection of larval food plants by females, this type of variation can have profound effects on the course of speciation, especially in species that exist as isolated colonies as does the Common Blue.

This also means that all potential food plants acceptable for larval development may not be utilized. Caterpillars use olfactory and taste receptors to distinguish the plant chemicals emanating between the epidermal cells and from within the epidermal pores, or stomata, on the leaf surface. However, Dethier (1975) points out that the olfactory organs probably do not "see" the leaf as biochemically uniform, and gustatory organs receive different signals from different parts of the leaves, as well as from leaves of different ages. Thus, compounds that stimulate feeding may not be present in the same concentrations within leaves or between leaves for a variety of reasons.

For example, the Monarch butterfly (*Danaus plexippus*) oviposits on only a very small percentage of available milkweed plants, and new growth is preferred both by ovipositing females and early instars (Borkin 1982). In addition, Cohen and Brower (1982) found that more eggs were laid on larger milkweed plants. Yet, the number of eggs per plant and larval success rates were not related to the amount of cardiac glycoside concentrations within the plants.

Interestingly, both Borkin (1982) and Cohen and Brower (1982) report that only 4 to 12 percent of the eggs laid survived to the fifth instar, despite the presence of the vertebrate toxins (see chap. 6). Larger plants may be chosen because they afford more protection from environmental stress—heat, direct sunlight, rain, and hail—as well as a larger canopy to protect eggs and larvae from invertebrate predators and parasitoids. Monarch caterpillars may gain some protection from vertebrates with their warning stripes of black, yellow/orange, and white, but this color pattern does little to deter attack from other arthropods.

Monarch larvae can also apparently regulate to some extent the uptake and storage of cardiac glycosides (see chap. 6). Those reared on plants low in cardenolide concentrations sequester relatively more of the toxin per gram consumed than those fed on plants with high concentrations (Brower, Seiber, Nelson, Lynch, and Tuskes 1982). Borkin (1982) discovered that individual Monarch larvae often move between food plants during development, even though the initial plants are not overly defoliated. Why should a larva leave its food plant and expose itself to unnecessary danger? Perhaps it is for thermoregulatory reasons since most movement is recorded at midday; perhaps it is to escape parasites and parasitoids; but it is most likely because there is some physiological change that occurs in the original food plant that cannot yet be detected by us (Borkin 1982).

Plants suffering from long-term predation by specific groups of butterflies may evolve a variety of defensive structures such as hair-like trichomes, or defensive plant chemicals—the so-called second-ary plant compounds or allelochemicals. There are many different classes of defensive compounds, including tannins, phenols, and even a group of chemicals that releases cyanide when hydrolyzed by certain enzymes. Still another group of allelochemicals is the al-kaloids—nitrogen-containing substances that are produced by a number of distantly related vascular plants, and include compounds like quinine, strychnine, and nicotine (Bower, personal communi-cation).

Some alkaloids can kill mammals outright, deform their develop-ing embryos, or induce abortions. Many plants produce and store these and related toxic compounds in special cellular compart-ments, the vacuoles, presumably as a defense against the generalist-type herbivores. Generalists apparently do not or cannot detect the presence of these toxins beforehand, and thus their larvae are usu-ally susceptible to their actions. In some cases, generalist herbivores may detect alkaloids in the plants prior to oviposition, and thereby avoid such plants.

Dolinger and his colleagues (1973) investigated the predation pat-tern of the Silvery Blue butterfly (*Glaucopsyche lygdamus*) on popu-lations of perennial lupines in Colorado. The herbaceous lupines produce different amounts and kinds of alkaloids in their flowers. The tiny Silvery Blue larvae, which complete their development on a single plant, are apparently not affected by these flower alkaloids. Perhaps they excrete them, metabolize them to nontoxic by-prod-ucts, or even sequester them for their own protection.

The researchers found that lupine populations in areas ecologi-cally unfavorable to *G. lygdamus* had few alkaloids in their flowers, and these showed little variability, whereas populations susceptible to lycaenid predation had much higher concentrations of alkaloids. However, those plant populations with individuals containing variable mixtures of three or four alkaloids in their flowers suffered the least damage, while those containing a relatively invariant mixture of alkaloids suffered the most damage. It appears then, that plant populations composed of individuals whose chemistry varies both in the number and concentration of alkaloids have the best lupine defense against specialist herbivores such as the Silvery Blue (Dolinger, Ehrlich, Fitch, and Breedlove 1973).

A very thorough study of the relationships between butterflies and their food plants as mediated by a specific class of compounds was conducted by Chew (1975). She studied the food plant preferences of Colorado populations of the Western White (*Pieris occidentalis*) and the Mustard White (*P. napi macdunoughii*). Females of both species oviposit on members of the Cruciferae (mustard family), whose leaves contain compounds called glucosinolates and their related products of enzymatic hydrolysis, the isothiocyanates.

These compounds alone or together induce larvae to feed. The Mustard White develops equally well on four native species of crucifers, refuses one native species, and dies on the crucifer *Thlaspi arvense*, a now "naturalized" weedy species introduced from Europe. However, those reared on the native crucifer *Descurainia richardsonii* reached the pupal stage in the shortest period of time. Similarly, *P. occidentalis* develops equally well on the four native crucifers, but also dies on *Thlaspi arvense*. Again, development times varied but not pupal weights, indicating that all four plants were equally suited nutritionally for larval development, at least as far as survival is concerned (Chew 1975; Chew, personal communication).

Chew argues that selection in the Colorado populations favors rapid larval development, especially since early summer droughts can be severe enough to cause irreversible wilting of food plants, and also make larvae vulnerable to cold autumns (Chew 1975). But what about oviposition by both species of butterflies on *Thlaspi arvense*—a plant toxic to all early instars? Why, if selection is operating, do females continue to oviposit on this plant?

Chew observed the oviposition behaviors of both species under natural field conditions in an attempt to answer this question. Both species utilized all host plant species as nectar sources, but the Mustard White failed to oviposit on *Erysimum asperum*, a plant rejected even by starving larvae, and *Cardamine cordifolia*, a crucifer associated with more shaded willow swales.

If ovipositing females could assess the suitability of their cruciferous hosts, then both should avoid *Erysimum asperum* (refused by larvae) and *Thlaspi arvense* (lethal to larvae). Yet, over 50 percent of the eggs from both species were placed on *Thlaspi arvense* at one test site, while many others were placed on rosettes—plants too small to support complete larval development. Generally, however, the oviposition preferences of the *Pieris* butterflies at each test site reflected the suitability of each plant species to support larval growth and development (Chew 1977). Thus, Chew points out that although oviposition is almost entirely on plants containing glucosinolates, the presence of glucosinolates is not always a reliable indicator of a suitable larval food plant.

The native crucifer avoided by larvae, *Erysimum asperum*, probably has coexisted with its pierid predators for thousands of years, at least since the last glacial period. Selection presumably has operated to engrave the message of unsuitability into the gene pool of the pierine species. But *Thlaspi arvense* was introduced only about one hundred years ago, and because it too contains glucosinolates, the correct stimuli are present for oviposition and feeding.

We might envision two possible results of this situation. A genetic combination may occur in some pierine offspring that allows larvae to tolerate and develop on the food plant, or the gene pool may undergo negative selection for hundreds, perhaps thousands, of years before those females that oviposit on *Thlaspi arvense* are selected out of the population. As of now, each female that oviposits on this lethal plant will contribute fewer genes on the average to the next breeding generation. Thus, if selection is operating, the pierids might either cease ovipositing on *Thlaspi arvense*, or evolve a tolerance for it in the early larval instars (Chew 1977).

Rodman and Chew (1980) later characterized the glucosinolate "profiles" of the Colorado Cruciferae and found that two naturalized weeds, including the lethal *Thlaspi arvense*, produce a presumed pierid attractant, allylglucosinolate. This glucosinolate is present also in the preferred species that allows the most rapid larval development, *Descurainia richardsonii*. Of twenty-two different glucosinolates recorded for these plants, up to eleven are found in the palatable species while only four are present in the lethal species.

However, *Descurainia richardsonii* was distinguished from all others by the presence of 3-butenyl-glucosinolate. Adults probably discriminate between the different crucifer species by "tasting" the presence of various glucosinolates with their tarsal chemoreceptors (e.g., see Fox 1966; Calvert 1974). The story of Colorado's *Pieris* butterflies and their cruciferous host plants is likely more complicated than what is related here. Perhaps other compounds, including attractant and/or deterrent compounds, as well as heretofore unknown physical stimuli are also involved in oviposition and feeding responses (Rodman and Chew 1980).

Chew's work is a good demonstration that butterfly behavior can be correlated with small variations within a single chemical class. Thus, when two plants contain similar variations, the butterflies treat them similarly. The *Descurainia* and the lethal *Thlaspi* are the only plants in Chew's pierine community that contain allyl- or 3-butenyl-glucosinolates—nothing else is structurally related.

Chew believes that the pierines mistake the lethal species for the good one because during their evolutionary (rather, coevolutionary) history this specific chemical profile has always been associated with good food plants. Also, this system suggests that plants exploited by a herbivore are under selection to become distinct from each other, so that the butterfly doesn't confuse them. Thus, coevolution between the pierine butterflies and their cruciferous hosts also involves interaction among the cruciferous plants (Chew, personal communication).

The rate of larval development also varies between closely related butterflies that utilize plants with different growth forms—herbs, shrubs, or trees. An excellent example of this phenomenon is reported by Scriber and Feeny (1979). They discovered that North American papilionid larvae grow faster and assimilate food more efficiently on herbaceous plants than on shrubs and trees. The high growth rates of the most precocious species are correlated to the water content of the food plant leaves. In general, the foliage of herbaceous plants has a higher water content than that of shrubs or trees.

The papilionid species exhibiting the slowest growth rates were typically polyphagous, including the Tiger Swallowtail (*Papilio glaucus*), whose larvae can be found on plants from at least eleven families. The Tiger Swallowtail—a true generalist—might be paying a metabolic price for its lack of food plant specificity—that is, slow growth. This may be because a generalist must have a greater metabolic investment to cover detoxification of food plant allelochemicals. Specialists such as the Black Swallowtail (*P. polyxenes asterius*), by contrast, feed on plants within a single family—the Umbelliferae (carrot family)—and thus hypothetically require fewer types of enzymes to detoxify any allelochemicals present.

The research of Scriber and Feeny does not really support this contention, and they also point out that it may not be water content per se that allows herbaceous feeders to develop faster than shrub and tree feeders, but other characteristics that correlate with increasing water content, such as increasing nitrogen but less fiber, which are known to affect the rate of larval development. Herbaceous plants generally have more nitrogen per gram and are not as tough as shrubs and trees (Slansky and Feeny 1977).

Scriber and Feeny (1979) suggest that shrub and tree leaves are more "apparent" through time to predators because these plant forms are long-lived, whereas many herbaceous species are short-lived and therefore less apparent through time and space. Because they are long-lived, shrubs and trees might invest more heavily in biochemical arsenals, whereas herbaceous species are more ephemeral in the habitat and do not require such defense excesses. This might very well explain why tree leaves are "tougher" all around—as a consequence of adaptation to inhibit herbivore damage.

However, herbaceous plants often sprout earlier in the spring—particularly woodland species—and thus they are quite available to herbivorous insects. In addition, other more primitive long-lived herbaceous plants such as the bracken fern (*Pteridium aquilinum*) have a diverse chemical arsenal that varies in quality and quantity throughout the growing season. Yet they still suffer heavy predation, even after the leaves toughen and become less palatable throughout the summer (Douglas 1983). Also, many tree-feeding papilionids choose newly emerged leaves on the smallest trees (less than 2 meters [6 feet] in height) for oviposition. We know little of how the chemical composition of their leaves change over the growing season, nor do we know how these changes affect their palatability to butterfly larvae.

Thus, until we learn where each papilionid species lays the majority of its eggs (e.g., old growth or new growth), and the characteristics of the leaves utilized by the larvae in the field by different broods, we can say little more about the developmental rates of larvae and the correlations to food plant growth form. Interestingly, though, Futuyama (1976) and Slansky (1976) observed that species that feed on the leaves of trees and shrubs are less restricted to single families of host plants than species whose larvae feed on herbaceous plants.

Even specialists such as the Black Swallowtail have oviposited on the "wrong" family of food plant. Yet such oviposition mistakes are often very successful. I have reared lepidopterans as diverse as the Gulf Fritillary (*Agraulis vanillae*), the Regal Fritillary (*Speyeria idalia*), and the saturniid moth *Rothschildia forbesi* on an artificial diet whose major component was pulverized lima beans. Larvae occasionally can be started on such artificial diets, but they have a better chance of successfully completing metamorphosis if they are transplanted at later instars onto a normal host plant. Larvae fed on artificial diets may have extended developmental times and the adults may be stunted.

It appears, then, that the selection and subsequent successful development of a butterfly on a new food plant involves several different components. First, the adult females must oviposit on the food plant regularly by mistake, for whatever reason. Second, the alternate food plant must provide physical or biochemical inducements for the newly hatched larvae to feed.

If chemical deterrents, toxins, or other forms of allelochemicals are absent, chances are that the nutritional requirements of the larvae will be satisfied. The fact that many species can develop successfully on many different species of nonhost plants if they are somehow given the opportunity, or "tricked" by an experimenter, does not diminish the complexity of coevolutionary give-and-take that has occurred between butterflies and their food plants over millions of years. It simply means that selection has operated to partition the natural habitat in a variety of ways and for a number of different reasons at different points in time. However, extreme mutualistic dependencies between butterflies and their host plants are rare.

The potential number of plant-butterfly interactions is far greater than the number of plant species or the number of butterfly species taken alone (Benson, Brown, and Gilbert 1976). As larvae, butterflies are typically predators and plants are their prey (Breedlove and Ehrlich 1971). As adults, though, butterflies may be mutualists that increase the reproductive success of their flowering plant hosts (e.g., Levin and Berube 1972), or the pollen from these plants may conversely enhance the reproductive output of the pollinating butterflies (Dunlap-Pianka 1979; Gilbert 1972). In at least one case, adult butterflies (ithomiines) are believed to become unpalatable through adult feeding, not from their solanaceous larval host plants (Gilbert, personal communication).

If Ehrlich and Raven (1965) are correct, plants have diversified (radiated) during times of escape from their butterfly predators. But how does plant chemistry affect the evolution of host specificity in butterflies? Smiley (1978a, 1978b) found that larval growth rates of *Heliconius* butterflies do not closely parallel their choice of host plants in the field. This indicates that other factors besides host plant chemistry are important in host plant selection.

Rausher, Mackay, and Singer (1981) determined that large isolated plants at one mid-elevation montane site in California received more clusters of eggs by ovipositing Edith's Checkerspot (*Euphydryas editha*) than would be expected from these plants' numerical representation in the population. They suggest that females may prefer large, isolated plants, and that acceptance of a host plant for oviposition is probably not a random event—that is, there is some choice by the females involved.

This would make sense since Rausher (1978) determined that the Pipevine Swallowtail female (*Battus philenor*) tracks changes in relative host plant suitability for larval growth. The two host plants involved—*Aristolochia serpentaria* (narrow leaves) and *Aristolochia reticulata* (wide leaves)—differ considerably from each other. Differences in search modes for these plants possibly reflect underlying differences in the females' response to leaf shape, but this probably does not mean they form a true "search image" for a given leaf shape (Rausher and Papaj 1983a, 1983b).

Differences in search modes may explain why ovipositing females of the Pipevine Swallowtail will alight on many plants with leaf shapes similar to their host plants, and determine their suitability as hosts only after drumming the plant with their foretarsi

(Rausher 1978, 1980; see also chap. 4). Interestingly, the predominant search mode in Rausher's Texas population of *Battus philenor* changes seasonally from broad-leaved in the early spring to narrow-leaved in the late spring, yet this switch is not correlated with any change in abundance of the two hosts (Rausher 1978, 1981b).

Leaf shape is likely the primary cue for initiating the alighting response of females (Rausher 1981b). Perhaps apostatic selection exerted by coevolved herbivores has produced this leaf shape diversity in these two closely related species of *Aristolochia*. The Pipevine Swallowtail may switch to *A. serpentaria* from *A. reticulata* because *A. reticulata* leaves become tough and inedible by late spring, whereas *A. serpentaria* are edible throughout the season. However, on any given day females having both types of search images (in the loose sense of the term) can be found in the population. Because females will not lay eggs on tough leaves, the rate of oviposition on *A. reticulata* declines as spring progresses (Rausher 1980, 1981b).

In a similar study, Stanton (1984) observed the short-term learning and the searching accuracy of three species of *Colias* butterflies. She determined that females partitioned their time into periods of searching for oviposition plants and periods of visiting flowers for nectar. Eggs were laid most frequently on legume host plants, although females also landed quite often on nonlegume species, especially those that superficially resembled suitable host plants.

Stanton also discovered that the average frequency of "landing errors" decreased over the course of an egg-laying flight, suggesting that females learned to identify host plants more accurately on the basis of experience. Landing accuracy was low after periods of nectar feeding, perhaps because there was a trade-off between the two searching modes. She suggests two hypotheses: (1) the partitioning of time into oviposition and nectaring modes enhances overall searching efficiency, and/or (2) experience and searching specialization may increase searching accuracy in invertebrates, as they do in vertebrates.

If females can recognize leaf shape and leaf pubescence, perhaps they can also assess the number of eggs already laid on a given host plant. Several studies support this hypothesis. For example, Rausher (1979) determined that ovipositing females of *Battus philenor* can detect the presence of eggs laid by other females on their host plants.

Females may spend several minutes searching for eggs, to her off-spring's benefit, since survivorship from egg to adult is likely to be lower on plants that are already overtaxed by too many larvae.

Females that locate eggs on potential host plants are inhibited from laying more eggs on the plant. Because small larvae cannot move easily from plant to plant without the danger of starvation or predation, host plant choice exercised by female butterflies can greatly influence her success at producing offspring. Rothschild and Shoonhoven (1977) have reported similar assessment of "egg load" on host plants by the European pierid *Pieris brassicae*, and Shapiro (1980d, 1981a, 1981b) has noted *egg-load assessment* in North American pierines. In these cases, the function of discrimination against host plants may be to prevent overloading the plant.

The symbiotic system of *Battus philenor* and its *Aristolochia* host plants in Texas continues to be unraveled by Rausher and his colleagues (e.g., Rausher and Papaj 1983b). It is clear now that the butterflies use visual cues to search for plants with specific leaf shapes, and then test these potential food plants with chemical re-ceptors on their foretarsi to make the proper identification. Follow-ing this action, they search the plant to assess egg load, and only then proceed to oviposit. This is a very sophisticated coevolutionary system involving plants and butterflies, but for sheer intricacy of plant-butterfly coevolution, one has to turn to the unsurpassed *Heli-conius-Passiflora-Anguria* system.

This system involves approximately forty-five species of nectar-and-pollen-feeding *Heliconius* butterflies, their larval food plants, the passionflowers (Passifloraceae), and their adult food sources, the *Anguria* and *Gurania* ("rainforest cucumbers" in the family Cucur-bitae) vines that they pollinate. The story was and continues to be unraveled largely through the efforts of Lawrence Gilbert of the University of Texas. The following is a composite story constructed from many years of research.

The larvae of *Heliconius* butterflies feed entirely on *Passiflora* and several closely related genera (Benson, Brown, and Gilbert 1976). Taxonomists place some three-hundred-fifty-odd species of plants within the *Passiflora* genus, but in any given locality there are usually no more than ten species. The species of each locality typically exhibit a striking diversity of intra- and interspecific leaf shape, almost unheard of for tropical plants.

Heliconius larvae can extensively damage their specific host *Passiflora* plant, thereby reducing the plant's ability to flower and fruit. Like the *Asclepias* (milkweeds), most *Passiflora* have a biochemical arsenal including cyanide-producing compounds and alkaloids that apparently keep most generalist herbivores at bay. Some of these toxins may be sequestered by *Heliconius* larvae, rendering them unpalatable to some unknown degree, but this has not been experimentally verified (Gilbert, personal communication).

Each species of *Passiflora* differs slightly in leaf chemistry, and may even smell differently. Many have extrafloral nectar glands that secrete a sugary solution attractive to ants and other insects. The nectaries, in effect, serve as a second line of defense (Ehrlich and Gilbert 1973) by providing the plants with ants that may deter oviposition or attack young larvae.

Some *Passiflora* species, such as the widespread *P. adenopoda*, have evolved hooklike trichomes—sharp, curved plant hairs—that serves as a sort of mechanical defense (Gilbert 1971). The trichomes cling to and puncture the cuticle around the prolegs, immobilizing the caterpillar. Death results from extensive loss of haemolymph and starvation. Gilbert hypothesizes that these hooked trichomes are a relatively recent evolutionary innovation, and probably developed long after biochemical defenses targeted at specialists like heliconiine butterflies failed.

Still, Rathcke and Poole (1975) have documented an unusual case where an ithomiid butterfly, *Mechanitis isthmia*, has avoided trichomes on its solanaceous hosts by feeding gregariously on the unprotected leaf edges, suspended from a silk scaffold spun over the tops of the trichomes. Gilbert and his colleagues have also determined recently that the Zebra butterfly (*Heliconius charitonius*) and the heliconiine butterfly *Dione moneta* can also use *Passiflora* species with hooked trichomes (Gilbert, personal communication).

But back to our *Heliconius* story. The distasteful heliconiines possess an amazing behavioral repertoire. The slow-flying adults are aerodynamically proficient—they can hover, fly backward, and flit upside down, all with graceful ease. Their huge compound eyes and optic lobes endow them with a broad visual spectrum as well as keen eyesight—good enough, as we have seen, to navigate through the guide lines supporting the webs of orb-weaving spiders. They are also keen enough to recognize leaf shape and even landmarks within their local home ranges. The long-lived adults roost gregariously (chap. 5) at night with a high degree of faithfulness to a particular roost, so that whole families and extended lineages may occupy a single roost.

The *Heliconius* butterflies also exhibit complex sexual and reproductive behaviors. Males in one group of the genus (including *H. erato, charitonius, sara, sapho,* and others) are attracted by pheromones to female pupae with pharate adults. Here they sit and await eclosion of the females, although males of some species are known to "rape" the pupae before eclosion, at least under laboratory conditions. In fact, several species apparently rape the pupae of closely related species. The extremely angular and leaflike pupae are also interesting because they exhibit a color polyphenism, and, like the lycaenids, they can stridulate and probably communicate with adults. Males have been shown to contribute nitrogen sources as well as antiaphrodisiac pheromones to females during mating (see chap. 7).

Ovipositing females are believed to learn and associate the leaf shapes of their preferred host plants with their chemistry—this may sometimes cause the visually searching females to investigate non-host plants with similar leaf shapes. Females carefully inspect a host plant before laying eggs, perhaps, as Gilbert (1975) suggests, to avoid ovipositing on plants harboring egg parasitoids and predators—or even other *Heliconius* eggs and larvae.

In fact, there are a number of *Passiflora* species that have evolved fake yellow eggs that appear at different locations—stipule tips or tendrils, for example (Benson, Brown, and Gilbert 1976). Searching females are less likely to oviposit on vines containing real eggs or the egg mimics, at least in the laboratory (Williams and Gilbert 1981), but Shapiro (1980d) has documented egg-load assessment by pierines in the field, so it would not be surprising to find that *Heliconius* females do likewise.

There are other traits of *Passiflora* that possibly evolved with increasing "transparency" to *Heliconius* butterflies. Some species of *Passiflora* have stipules resembling tendrils that are shed with the egg of any species that oviposits on new tendrils. By contrast, plants with stipules bearing egg mimics are not deciduous.

Some *Passiflora* species have tough old leaves with undigestible compounds that make the leaves unsuitable for larvae, or leaves that are shed in response to larval damage. Still, as we have seen, others have extrafloral nectaries that secrete a sugary solution that attracts mutualistic defenders such as predaceous ants (Gilbert 1979). The *Passiflora* plant morphology, their variation in leaf shape, the searching behavior of ovipositing butterflies, and the existence of egg mimics taken together supply convincing evidence that extensive coevolution between *Heliconius* butterflies and their larval host plants continues to take place (Gilbert 1982).

Many but not all species of *Heliconius* are specialists on a given set of *Passiflora* species (Smiley 1978a, 1978b; Gilbert and Smiley 1978), but a given locality usually contains no more than ten species of heliconiines, each with a corresponding primary food plant (Gilbert 1975). Excellent vision and long reproductive lives allow *Heliconius* butterflies to become very familiar with landmarks around the nocturnal roosts. This familiarity allows them to locate *Gurania* and *Anguria* plants, the primary pollen sources to which they owe their extended life and reproductive capacity.

These two Neotropical plant genera at first appeared to have skewed sex ratios in which male plants typically outnumbered female plants by as many as ten to one. Now it is known that although male flowers predominate, these protandrous plants are delayed "sex-switchers," and not dioecious (Gilbert, personal communication). The pollen-producing flowers of the male plants are produced one by one on a long trailing peduncle. There may be more than one peduncle per plant, each of which typically produces a single male blossom each day, or every other day. And although the male flowers last only one day, the plants may flower continuously for up to three years, and thus present a very reliable nectar and pollen resource for the adults (Gilbert 1975).

Females of one well-studied species, *Heliconius ethilla*, forage on the flowers in the early morning between 5:30 and 6:30, but other species in other habitats show slightly different patterns. Nonetheless, early foraging by at least some species is common across many rainforest habitats. By foraging early the *Heliconius* butterflies obtain most of the pollen from each day's flowers.

Gilbert calculates that a single butterfly load of pollen from *Anguria* flowers is sufficient to produce five eggs—about the daily production level of most heliconiine species. The foraging patterns of *Heliconius* females are unique in yet another way. Because they rely almost entirely on *Anguria* and *Gurania* vines for pollen sources, they have evolved a method of efficiently extracting pollen from the same plants each day. This method, called trap-lining, allows females to visit the plants in a regular order, thereby minimizing energy expenditure and exposure to natural hazards.

The female *Anguria* flowers ultimately are pollinated by *Heliconius* butterflies (even though they cannot provide pollen), probably because they mimic the male flowers in shape and color. Nectar production typically occurs during the daylight hours after pollen production has ceased, thereby reinforcing the flower-visiting patterns of trap-lining females—and males. Thus, selection has operated to increase foraging efficiency on the cucurbit vines to the mutual benefit of both groups of organisms.

Visiting the same pollen-bearing plants in nearly the same sequence each day is likely an adaptation to improve the efficiency of pollen collection. Communal roosting, learning capacity, and a well-developed visual system also are traits in *Heliconius* that are selectively advantageous and improve pollen and nectar collection. Gilbert hypothesizes that gregarious roosting might have evolved when young butterflies followed older butterflies on their pollen collecting routes, then congregated together at night, but this remains to be verified.

It may be that finding a safe roost is just as critical (Gilbert, personal communication). Because the butterflies are unpalatable, a communal roost serves to reinforce the effectiveness of the "bad" advertising. In some locations (e.g., Trinidad) the various species of *Anguria* may divide up the period during which nectar production is maximum, some peaking before noon, others not peaking until late in the afternoon (Gilbert 1975).

Although *Heliconius* females use visual cues in addition to chemical cues to increase their efficiency of finding suitable host plants (Gilbert 1982), one relatively direct way of escaping oviposition—hence predation by larvae—is to evolve a leaf shape that is different from that of the stereotypical food plant, or one that matches some other nonfood plant species. Gilbert hypothesizes that female *Heliconius* are thus "agents of selection" on *Passiflora* shape (Gilbert 1982).

In fact, there is an unusually high diversity of *Passiflora* leaf shapes in most localities, regardless of the particular *Heliconius* species involved. If two local species have similar leaf shapes, they differ in other characteristics such as hairiness or tendril shape. And, in many cases, leaf shapes of *Passiflora* species have converged on those of non-*Passiflora* species unsuitable to *Heliconius*—suggesting a type of plant mimicry.

Thus, the number of *Passiflora* species, and the number of leaf shapes, typically correspond to the number of *Heliconius* species in a given locale. If, as Gilbert (1975) hypothesizes, the degree of shape discrimination by *Heliconius* butterflies places an upper limit on the number of heliconiine species that can occupy a given community, then it follows that *Heliconius* species can indirectly limit their own local diversity. If this is true, it would make the *Anguria-Heliconius-Passiflora* complex one of the most intricate coevolutionary systems ever discovered. The five major recognizable radiations of heliconiines seem to be able to coexist today because of fundamental differences in the manner in which they partition their host plants.

Postscript

There is no greater excitement in life than discovering something that no one knew before. While it may seem that the butterflies are an overstudied group of organisms, nothing could be further from the truth. I look forward to the day when I can revise this entire book—correcting and expanding the information we know about butterflies. To do this will require the input of amateurs and professionals alike. There is always room for experimentation, and we should never assume that we have all the answers to the complicated questions concerning the biology of butterflies.

Reams of research information are available on many butterflies such as the Monarch (*Danaus plexippus*) and the Cabbage White (*Pieris rapae*), but reams more remain to be written on all aspects of their behavior, physiology, and ecology. The field for original research at all levels is literally open ended. The types of research to be accomplished are limited only by the human imagination. I encourage everyone interested in butterflies to make their study more than a pastime or professional endeavor—make it your passion. We have barely scratched the surface, and we have a very long way to go.

Appendixes

APPENDIX A

Geologic Time Scale

ERA	PERIOD	EPOCH	TIME – in millions of years ago	HEIGHTS BELOW ARE PROPORTIONAL TO TIME · MILLIONS OF YEARS AGO
		HEIGHTS BELOW ARE NOT PROPORTIONAL TO TIME		
CENOZOIC	QUATERNARY	RECENT	— 0.004	CENOZOIC — 63 — / MESOZOIC 220 —
		PLEISTOCENE	— 0.5 - 2.0	PALEOZOIC
	TERTIARY	PLIOCENE	— 13 ± 1	600 —
		MIOCENE	— 25 ± 1	
		OLIGOCENE	— 36 ± 2	
		EOCENE	— 58 ± 2	
		PALEOCENE	— 63 ± 2 —	
MESOZOIC	CRETACEOUS		— 135 ± 5	
	JURASSIC		— 180 ± 5	
	TRIASSIC		— 220 ± 10 —	
PALEOZOIC	PERMIAN		— 280 ± 10	
	PENNSYLVANIAN		— 310 ± 10	
	MISSISSIPPIAN		— 345 ± 10	
	DEVONIAN		— 405 ± 10	
	SILURIAN		— 425 ± 10	
	ORDOVICIAN		— 500 ± 10	Precambrian (ARCHEOZOIC and PROTEROZOIC) Eras
	CAMBRIAN		— 600 ± 20 —	
	"PRECAMBRIAN"			*OLDEST RADIOGENIC ROCK DATE* — — 3500 —
				EARTH ORIGIN — 4500 - 5000 —

From John A. Dorr, Jr., and Donald F. Eschman, *Geology of Michigan*. Ann Arbor: University of Michigan Press, 1970.

APPENDIX B

List of Some American Butterflies Used in Scientific Research

COMMON NAME	SCIENTIFIC NAME
Alaskan Western White	Pieris occidentalis nelsoni
Anicia Checkerspot	Euphydryas anicia
Anise Swallowtail	Papilio zelicaon
Baltimore Checkerspot	Euphydryas phaeton
Banded Hairstreak	Satyrium calanus
Bay Checkerspot	Euphydryas editha bayensis
Behr's Sulphur	Colias behrii
Black Swallowtail	Papilio polyxenes asterius
Bordered Patch	Chlosyne lacinia
Bronze Copper	Lycaena hyllus
Buckeye	Precis coenia
Cabbage White	Pieris rapae
California Arctic	Oeneis ivallda
Canada Arctic	Oeneis macounii
Checkered White	Pieris protodice
Cloudless Sulphur	Phoebis sennae
Comma	Polygonia comma
Common Blue	Plebejus icarioides
Common Sulphur	Colias philodice
Compton Tortoise Shell	Nymphalis vau-album
Coral Hairstreak	Harkenclenus titus
Dainty Sulphur	Nathalis iole
Diana Fritillary	Speyeria diana
Dog-face	Colias cesonia
Dorcas Copper	Lycaena dorcas
Edith's Checkerspot	Euphydryas editha
Fairy Yellow	Eurema daira
Goatweed Butterfly	Anaea andria
Great Copper	Lycaena xanthoides
Great Southern White	Ascia monuste

COMMON NAME	SCIENTIFIC NAME
Great Spangled Fritillary	Speyeria cybele
Gulf Fritillary	Agraulis vanillae
Hackberry	Asterocampa leilia
Harvester	Feniseca tarquinius
Johnson's Hairstreak	Callophrys johnsoni
Julia	Dryas julia
Jutta Arctic	Oeneis jutta
Little Yellow	Eurema lisa
Mead's Sulphur	Colias meadii
Milbert's Tortoise Shell	Nymphalis milberti
Monarch	Danaus plexippus
Mourning Cloak	Nymphalis antiopa
Mustard White	Pieris napi
Nevada Arctic	Oeneis nevadensis
Old World Swallowtail	Papilio machaon
Orange Sulphur	Colias eurytheme
Painted Lady	Vanessa cardui
Pearl Crescent	Phyciodes tharos
Pearly Eye	Lethe portlandia
Pink-edged Sulphur	Colias interior
Pipevine Swallowtail	Battus philenor
Polaris Fritillary	Boloria polaris stellata
Purplish Copper	Lycaena helloides
Queen Alexandra's Sulphur	Colias alexandra
Queen	Danaus gilippus
Question Mark	Polygonia interrogationis
Reakirt's Blue	Hemiargus isola
Red Admiral	Vanessa atalanta
Red-spotted Purple	Limenitis arthemis astyanax
Regal Fritillary	Speyeria idalia
Sara Orange Tip	Anthocaris sara
Scudder's Sulphur	Colias scudderi
Silver-bordered Fritillary	Boloria selene
Silvery Blue	Glaucopsyche lygdamus
Sleepy Orange	Eurema nicippe
Spicebush Swallowtail	Papilio troilus
Spring Azure	Celastrina argiolus pseudargiolus
Tiger Swallowtail	Papilio glaucus
Viceroy	Limenitis archippus
West Virginia White	Pieris virginiensis
Zebra	Heliconius charitonius

Glossary

Abdomen The third and posterior body division of the insect body.

Aberration An insect with traits that are unusual for the species, such as unusual scales or wing color patterns.

Aedeagus The penis of the male.

Accessory pulsatile organs Minute pumping organs that assist in the circulation of haemolymph throughout the body.

Aeropyles Minute breathing pores in the egg shell that provide for the exchange of gases.

Aestival morph The summer form of a polyphenic species.

Alleles Alternate forms of a gene.

Anal cerci A pair of segmented appendages that extend from the posterior area of the abdomen.

Analogous A term describing structures that are similar in function but different in embryonic origin.

Ancestral A term describing a structure or group of organisms that appear to be more primitive in the evolutionary sense than other related structures or related groups of organisms.

Androconial patch A distinct patch containing androconial scales.

Androconial scales Highly modified scales of male butterflies that secrete a variety of chemical substances used in "seducing" or stimulating females.

Anal veins A series of three veins confined to the extreme base of the wing.

Antagonistic muscles In insects, sets of muscles whose actions oppose one another.

Antennae A pair of segmented sensory appendages located above the mouthparts on the head.

Aorta The anterior vessel of the butterfly "heart."

Aphrodisiac pheromones Pheromones emitted by males to calm and make a female ready for copulation.

Arolium The central, terminal structure of the distitarsus, flanked by the paronychia and the claws.

Asynchronous muscles Muscles in which a single neural impulse produces many muscle contraction-relaxation cycles.

Atrium The opening that leads to the ductus bursae.

Automimics Edible mimics of distasteful model butterflies within a single species (e.g., *Danaus plexippus*).

Balanced seasonal polymorphism A polyphenic species that exhibits several morphs throughout the year, each of which is under selective pressure.

Basal area The area near the base of the wing.

Batesian mimicry A mimicry system in which at least one distasteful model is mimicked by one or more edible species.

Battledore scales Racket-shaped androconial scales.

Biennial Appearing every other year.

Bilateral gynandromorph An aberrant butterfly that exhibits male traits on one side of the body and wings and female traits on the other side.

Biological species definition Species are distinct groups of organisms that do not interbreed with other such groups.

Bivoltine A species with two relatively discrete broods each year.

Body basking A basking position in which the wings are opened just enough to expose the dorsal surface of the body to sunlight.

Brain hormone A hormone that is cyclically released by the neurosecretory cells in response to external and internal stimuli. It mediates the production of other hormones involved in the molting process.

Bursa copulatrix The area of the female genitalia that receives and stores the spermatophore.

Cardiac glycosides A class of compounds produced by plants that may produce an emetic response and heart rhythm irregularities in vertebrates; used as an anti-predator defense by some insects.

Chaetosemata A pair of adult organs located on the top of the head whose function is probably sensory in nature.

Character states The specific traits or characteristics (e.g., "green" or "blue") of a character (e.g., "eyespots").

Chitin A nitrogenous polysaccharide found in the exoskeleton of insects and responsible for many of its distinctive properties.

Chorion That layer of the eggshell derived from secretions of the ovarian follicular cells.

Chrysalid The chrysalis or pupal state.

Chrysalis The metamorphic instar (pupa) of the butterfly.

Cibarium The cavity between the hypopharynx and the epipharynx.

Claspers The paired valvae of the male which aid in copulation.

Coevolution A type of "community evolution" in which there are selective interactions between two or more species with close ecological relationships.

Colleterial glands Special accessory glands in the female that secrete an adhesive cement around the egg during oviposition.

Compound eye An eye composed of many individual ommatidia.

Conspecific Referring to individuals belonging to a single species.

Coprophilous Referring to butterfly adults that use excrement as a food resource.

Corneal lens The transparent, cuticular part of the ommatidia.

Corpora allata Small paired lobes immediately behind the brain that secrete specific hormones.

Corpora cardiaca A pair of small storage bodies lying adjacent to the corpora allata.

Corpus bursae The main part of the bursa copulatrix as distinguished from the opening or ostium bursae.

Costal vein (costa) A longitudinal wing vein forming the anterior margin of the butterfly wing.

Coxa The basal segment of an insect leg.

Cremaster A structure bearing hooked processes at the posterior end of the pupa, used to attach the pupa to a silken support pad.

Crepuscular Referring to flight during the twilight.

Crochets Hooked spines on the plantar surface of lepidopteran larval prolegs.

Crop An enlarged area of an insect's foregut, lying just behind the esophagus.

Cross veins   Veins that bridge or connect longitudinal veins.

Crypsis   Referring to the ability of an organism to conceal or camouflage itself through color, pattern, or body structures.

Cubitus   A two-branched vein in the forewing and the hind wing.

Diapause   A period of arrested development during which growth, tissue differentiation and metamorphosis proceed at negligible rates.

Dimorphism   Two distinct forms.

Discal Cell   The enlarged cell in the central area of the wing.

Disruptive coloration   A color pattern that breaks up the body outline so that the butterfly does not stand out against its background.

Distitarsus   The fifth and terminal subsegment of the tarsus.

Dorsal basking   Basking in which the wings are fully open or nearly so to incoming sunlight.

Dorsal longitudinal muscles   The muscles running dorsally and lengthwise between segments.

Dorsal-ventral muscles   The muscles running vertically within segments.

Ductus bursae   The membranous tube leading to the corpus bursae.

Ductus seminalis   The tube connecting the common oviduct to the bursa copulatrix.

Ecdysis   The actual shedding of the old cuticle during the molting process.

Ecdysone   A hormone involved in the molting process.

Eclosion   The hatching of the larva from the egg; the emergence of the imago from the chrysalis.

Ecological range expansion   A population expansion into areas that a species was formerly unable to inhabit, usually because of genetic (physiological) adaptations within the population.

Egg-load assessment   The ability of gravid females to assess the density of conspecific eggs on a given food plant.

Escape response   Directed flight away from a predator, often toward the sun.

Exons   Those sections of DNA whose products are ultimately spliced together to form proteins.

Exoskeleton   The external cuticular skeleton.

Fat bodies   Storage sacs for fats that are particularly enlarged in migratory or diapausing butterflies.

Femur   The third leg segment located between the trochanter and the tibia.

Fitness   A measure of the relative reproductive success of individuals within a population.

Flagellum   The whiplike antennal segments beyond the scape.

Form drag   Drag which varies according to the shape the butterfly presents to the wind.

Frenulum   A type of wing coupling mechanism in which a series of bristles from the top of the hind wing fit into a slip on the bottom of the forewing (e.g., a frenate coupling mechanism).

Friction area   The area where the forewing overlaps the hind wing.

Frons   A triangular sclerite in the front of the head.

Furca   A forked structure that guides the penis into the female atrium during copulation.

Galeae   The outer lobes of the maxillae, forming the proboscis.

Gene flow   The exchange of alleles between populations due to interpopulational matings.

Genetic drift   The random increase or decrease in the relative abundance of different alleles, particularly effective in small, isolated populations.

Haemolymph   Insect "blood" that fills the haemocoel.

Hair pencil   A brush-like structure that dispenses pheromones in male butterflies (e.g., *Danaus plexippus*).

Hibernation   A winter dormancy (see diapause).

Hill-topping   The congregation of butterflies, especially males, above hill tops and elevated areas.

Homeothermic   Referring to the ability of an organism to regulate its body temperature within a relatively narrow range through physiological means.

Homologous   Referring to structures derived from similar embryonic tissues, but not necessarily having the same function.

Honey dew   A sugary solution secreted by special glands of lycaenid larvae.

Hydrostatic exoskeleton   Referring to the larval exoskeleton whose turgor is due to the fluid pressure within.

Hypopharynx   A soft, middle mouth part structure that lies in front of the labium and forms part of the proboscis. In the larval stage it forms part of the spinneret.

Imaginal discs   Undifferentiated embryonic cells of the epidermis which permit the metamorphic transformation to occur.

Imago   The adult.

Immature   Eggs, larvae, or pupae, but especially referring to the larvae.

Indirect muscles   The flight muscles used to change the shape of the thorax and indirectly operate the wings.

Instar   An immature larva between successive molts.

Interspecific   Between species.

Introns   The sections of a gene not translated directly into a polypeptide or protein.

Juvenile hormone   The hormone secreted by the corpora allata that maintains the juvenile characteristics and suppresses the development of adult characteristics.

Kin selection   An extended form of natural selection in which gene frequencies may be affected by close relatives who statistically share much of their genomes.

Labia   The lower, paired lip of the larval stage.

Labial palps   Small, knoblike projections on the labia.

Labial silk glands   The glands producing the silk during the larval stage.

Labrum   An upper lip lying above the clypeus.

Lateral basking   A basking position in which the wings are closed and presented ventrally to the sunlight.

Malpighian tubules   Excretory tubules that arise near the anterior end of the hindgut and extend into the body cavity.

Mandibles   The anterior pair of chewing mouthparts in the larval stage.

Maxillae   One of the paired mouthpart structures immediately posterior to the mandibles.

Maxillary palps   Small knoblike structures on the maxillae.

Meconium   The waste material ejected after the imago emerges from the chrysalis; often pigmented.

Medial vein (media)   The longitudinal vein between the radius and cubitus.

Mesothorax   The second thoracic segment lying between the prothorax and the metathorax.

Metamorphosis   A change in form during development.

Metathorax   The third and posterior segment of the thorax.

Micropyle   A small opening at the top of the egg where sperm can penetrate to fertilize the eggs.

Microtrichia   Minute sensory hairs on the wings.

Migration   A directed movement of individuals.

Molting   The complete physiological process of building a new cuticle and shedding the old cuticle.

Monomorphic mimicry   Mimicry in which both sexes of a palatable species mimic an unpalatable model.

Monophagy   A kind of host plant selection in which females of a given species confine their oviposition to a very narrow range of host plants.

Mud-puddling behavior   The assembly of adults, largely males, around mud puddle margins in search of sodium or nutrients.

Müllerian mimicry   Mimicry in which two or more distasteful species share similar wing patterns and sometimes behavior.

Multivoltine   Having three or more distinct broods in a year.

Muscular thermogenesis   A type of thermoregulation in which the body temperature is elevated and maintained by the synchronous contraction of the indirect flight muscles.

Myrmecophilous   Referring to the "ant-loving" behavior of certain lycaenid species.

Nectar   The sugary secretions of flowers, sometimes containing free amino acids and other nutrients.

Nudum   The scaleless part of the flagellum.

Nymphalid plan   A possible archetypal wing pattern from which nearly all butterfly wing patterns can be derived.

Ocelli   A simple eye.

Oligophagous   Referring to species that choose and can develop on several different groups of host plants.

Ommatidium   An individual unit of the compound eye.

Osmeterium   An eversible Y- or V-shaped organ located behind the head of papilionid larvae, normally concealed, but emitting a powerful, repugnant odor when the caterpillar is disturbed.

Ostia   Holes in the heart that admit haemolymph from the haemocoel.

Ovaries   The egg-producing organs of the female.

Oviduct   A tube through which the eggs pass from an ovary.

Oviposition   The act of laying eggs.

Paranotal lobe theory   A theory proposing that insect wings evolved from broad, flaplike structures present on the thorax of primitively flightless insects.

Parasitoids   Internal parasites that kill the host during the course of their development.

Paronychia   The padlike structures of the distitarsus that flank the claws.

Pars intercerebralis   A complex of neurosecretory cells and ordinary nerve cells that communicate with the corpora cardiaca.

Pedicel   An antennal segment after the scape and before the flagellum.

Pharate   Referring to a stage of metamorphosis still enclosed within the integument of a previous instar, and especially to adults that can be seen through the transparent skin of the chrysalis.

Phenology   The appearance and disappearance of a species throughout the course of the year.

Pheromones   Special chemical compounds secreted by individuals that have a behavioral influence on other members of that species.

Photoperiod   The length of day.

Phragmata   Platelike structures extending from the dorsal wall of the thorax and internally used for sites of muscle attachments.

Pleiotropic genes   Genes whose products have multiple effects within an organism.

Pleura   Lateral sclerites of the body.

Pleurite   A lateral or pleural sclerite.

Poikilothermic   Referring to organisms whose body temperature fluctuates directly with that of the ambient environment.

Pollenia   Pollen-bearing sacs.

Polymorphic   Having many different forms or morphs.

Polyphagous   Referring to species that can complete their development on many hosts belonging to diverse plant families.

Polyphenism   A polymorphism determined by seasonal changes in ambient conditions such as photoperiod or relative humidity, and not reflecting genetic differences among the phenotypes.

Proboscis   In butterflies, the coiled, springlike sucking tube derived from the extended galae of the maxillae.

Prolegs   Fleshy abdominal legs of caterpillars.

Prothoracic glands   The glands in the brain that secrete the molting hormone, ecdysone.

Prothorax   The first thoracic segment.

Pupa   The chrysalis.

Radius vein   The longitudinal vein between the subcosta and the media.

Reproductive diapause   Diapause in which the adults are active or in hibernation, but whose sexual organs remain in a nonreproductive state.

Rhabdom   A rodlike, light-sensitive structure found in the ommatidium of the compound eye.

Scales   Minute, overlapping and flattened structures derived from trichogen cells.

Scape   The basal segment of the antennae.

Sclerotization   The process of polymerization and cross bonding of protein and chitin that produces the hardened (tanned) exoskeleton.

Sensilla   Sensory cells; sometimes the components of larger sense organs.

Sex-limited mimicry   Mimicry in which only males or females of given species are involved as mimics.

Sex patch (brand)   A patch of modified scales that secrete or distribute sex pheromones.

Sexual dimorphism   A condition in which the sexes are differently patterned or pigmented.

Speciation   The process whereby one species evolves into two or more species.

Species   A reproductively isolated group of breeding organisms.

Spermatophore   The sperm-bearing package of the male.

Sphragis   A structure secreted onto the ventral part of the posterior abdomen of the female during copulation.

Spinneret   An organ that secretes silk.

Spiracles   The openings through which air enters the tracheae.

Sternite   Any sclerite in the sternal region of a segment.

Stridulatory organ   An organ consisting of a rasp and file that when rubbed together produce an audible sound.

Subcosta vein   A longitudinal vein immediately behind the costa.

Subspecies   A reproductively isolated or geographically defined population of a given species, with distinct characteristics.

Symbiosis   Referring to organisms that live together in close association.

Sympatry   The condition of occupying the same geographic location (e.g., sympatric species).

Synchronous muscles   Muscles in which a single neural impulse produces only one muscular contraction.

Tagma   A distinct body region (e.g., head, thorax, abdomen).

Tarsus   The last segment of the walking leg.

Tarsomeres   The subsegments of a tarsus, typically five in number.

Tergite   A tergal sclerite; the dorsal surface of an abdominal segment.

Thermoregulation   The process of actively or passively regulating the body temperature.

Thorax   The second tagma of the insect body.

Tibia   The fourth segment of the leg, lying between the femur and the tarsus.

Tormogen   An epidermal cell that develops into the socket for a scale or a hair.

Tracheae   The pipelike tubes carrying air into the interior of the insect.

Tracheole   An extremely fine trachea.

Trap-lining   The strategy of visiting the same food resources in the same order from day to day.

Trichogen   An epidermal cell that develops into a seta or hair.

Trichoid sensilla   Long, narrow sensory structures common on the antennae and other sensory areas of the exoskeleton.

Trochanter   The leg segment between the coxa and the femur.

Tubercle   A small nipplelike protrusion.

Univoltine   Having one brood per year.

Valvae   The male claspers of the genitalia used to grasp the exterior genitalia of the female.

Veins   Thick-walled hollow tubes that support the wing membrane and supply its cells with tracheae and nerve branches.

Venation   The pattern of veins that make up the structural support of the insect wing.

Ventral longitudinal muscles   The muscles running ventrally and lengthwise between segments.

Vernal morph   The spring form of a polyphenic species.

Vitelline membrane   An internal membrane derived from the cell wall of the egg, and lying beneath the chorion.

Walking legs   The true, segmented legs of insects.

Warning (aposematic) coloration   Coloration or wing patterns that serve to advertise unpalatability or danger to potential predators.

Winglets   The ancestral winglike expansions of ancient insects homologous to the wings of modern pterygotes.

Wing load   The amount of body weight carried per square unit area of wing surface. A measure of the weight that each square unit area of wing must carry during flight.

Bibliography

Adler, P. H. 1982. Soil- and puddle-visiting habits of moths. *J. Lep. Soc.* 36:161–73.

Adler, P. H., and D. L. Pearson. 1982. Why do male butterflies visit mud puddles? *Can. J. Zool.* 60:322–25.

Ae, S. A. 1957. Effects of photoperiod on *Colias eurytheme*. *Lepidop. News* 11:207–14.

———. 1958. Comparative studies of developmental rates, hibernation and food-plants in N. American *Colias* (Lep., Pieridae). *Am. Midl. Nat.* 60:84–96.

———. 1959. A study of hybrids in *Colias* (Lepidoptera, Pieridae). *Evolution* 13:64–88.

Aiello, A., and R. E. Silberglied. 1978a. Life history of *Dynastor darius* (Lepidoptera: Nymphalidae: Brassolinae) in Panama. *Psyche* 85:331–45.

———. 1978b. "Orange" bands, a simple recessive in *Anartia fatima* (Nymphalidae). *J. Lep. Soc.* 32:135–37.

Alexander, R. D., and W. L. Brown. 1963. Mating behavior and the origin of insect wings. *Occas. Papers Mus. Zool. Univ. Mich.* 628:1–19.

Arms, K., P. Feeny, and R. C. Lederhouse. 1974. Sodium: Stimulus for puddling behavior by tiger swallowtail butterflies, *Papilio glaucus. Science* 185:372–74.

Atkins, M. D. 1978. *Insects in perspective.* New York: Macmillan Publishing Co.

Atsatt, P. R. 1981a. Lycaenid butterflies and ants: Selection for enemy-free space. *Amer. Nat.* 118:638–54.

———. 1981b. Ant-dependent food plant selection by the mistletoe butterfly *Ogyris amaryllis* (Lycaenidae). *Oecologia* 48:60–63.

Austin, G. T. 1977. Notes on the behavior of *Asterocampa leilia* (Nymphalidae) in Southern Arizona. *J. Lep. Soc.* 31:111–18.

Baker, H. G., and I. Baker. 1975. Studies of nectar-constitution and pollinator-plant coevolution. In *Coevolution of animals and plants,* ed. L. E. Gilbert and P. H. Raven. Austin and London: University of Texas Press.

Baker, R. R. 1968. A possible method of evolution of the migratory habits in butterflies. *Phil. Trans. R. Soc. London* (B). 253:309–41.

———. 1970. Bird predation as a selective pressure on the immature stages of cabbage butterflies, *Pieris rapae* and *P. brassicae. J. Zool.* 162:43–59.

Barker, J. F., and W. S. Herman. 1976. Effect of photoperiod and temperature on reproduction of the monarch butterfly, *Danaus plexippus. J. Insect Physiol.* 22:1565–68.

Barth, R. 1937. Muscles and mechanisms of walking in caterpillars. *Zool. Fahrb.*, *Anat.* 62:507–66.

Beck, S. D. 1968. *Insect photoperiodism.* New York: Academic Press.

Benson, W. W. 1971. Evidence for the evolution of unpalatability through kin selection in the Heliconiinae (Lepidoptera). *Amer. Nat.* 105:213–26.

————. 1972. Natural selection for Müllerian mimicry in *Heliconius erato* in Costa Rica. *Science* 176:936–39.

————. 1978. Resource partitioning by passion vine butterflies. *Evolution* 32:493–518.

Benson, W. W., K. W. Brown, Jr., and L. E. Gilbert. 1976. Coevolution of plants and herbivores: Passion flower butterflies. *Evolution* 29:659–80.

Benson, W. W., and T. C. Emmel. 1973. Demography of gregariously roosting populations of the nymphaline butterfly *Marpesia berania* in Costa Rica. *Ecology* 54:326–35.

Berenbaum, M. R. 1983. Coumarins and caterpillars: A case for coevolution. *Evolution* 36:163–79.

Berenbaum, M., and P. Feeny. 1981. Toxicity of angular furancoumarins to swallowtail butterflies: Escalation in a coevolutionary arms race? *Science* 212:927–29.

Bernard, G. D. 1979. Red-absorbing visual pigment of butterflies. *Science* 203:1125–27.

Bernath, R. F. 1981. Pupal polymorphism in the cabbage butterfly, *Pieris rapae* L. (Lepidoptera: Pieridae). Ph.D. diss., Boston University.

————. 1982. Pupal polymorphism in the cabbage white butterfly *Pieris rapae* L. In *The evolutionary significance of insect polymorphism*, ed. M. W. Stock and A. C. Bartlett, 57–63. Proc. Symp. at Natl. Meeting of the E.S.A.

Blakley, N. R., and H. Dingle. 1978. Competition: Butterflies eliminate milkweed bugs from a Carribbean island. *Oecologia* 37:133–36.

Boggs, C. L., and L. E. Gilbert. 1979. Male contribution to egg production in butterflies: Evidence for transfer of nutrients at mating. *Science* 206:83–84.

Boggs, C. L., J. T. Smiley, and L. E. Gilbert. 1981. Patterns of pollen exploitation by *Heliconius* butterflies. *Oecologia* 48:284–89.

Borkin, S. S. 1982. Notes on shifting distribution patterns and survival of immature *Danaus plexippus* (Lepidoptera: Danaidae) on the food plant *Asclepias syriaca*. *Great Lakes Entomologist* 15:199–206.

Bowden, S. R. 1979. Subspecific variation in butterflies: Adaptation and dissected polymorphism in *Pieris (Artogeia)* (Pieridae). *J. Lep. Soc.* 33:77–111.

Bowers, M. D. 1978. Overwintering behavior in *Euphydryas phaeton* (Nymphalidae). *J. Lep. Soc.* 32:282–88.

Bowers, M. D., I. L. Brown, and D. Wheye. 1985. Bird predation as a selective agent in a butterfly population. *Evolution* 30:93–103.

Breedlove, D. E., and P. R. Ehrlich. 1971. Coevolution: Patterns of legume predation by a lycaenid butterfly. *Oecologia* 10:99–104.

Brower, J. V. Z. 1958a. Experimental studies of mimicry in some North American butterflies. I. The monarch *Danaus plexippus* and the viceroy, *Limenitis archippus*. *Evolution* 12:32–47.

———. 1958b. Experimental studies of mimicry in some North American butterflies. III. *Danaus gilippus berenice* and *Limenitis archippus floridensis*. *Evolution* 12:273–85.

———. 1958c. Experimental studies of mimicry in some North American butterflies. II. *Battus philenor* and *Papilio troilus, P. polyxenes,* and *P. glaucus*. *Evolution* 12:123–36.

Brower, L. P. 1961. Studies on the migration of the monarch butterfly. I. Breeding populations of *Danaus plexippus* and *D. gilippus berenice* in south central Florida. *Ecology* 42:76–83.

———. 1977. Monarch migration. *Nat. Hist.* 86:41–53.

Brower, L. P., and J. V. Z. Brower. 1964a. Birds, butterflies, and plant poisons: A study in ecological chemistry. *Zoologica* 48:65–84.

———. 1964b. Birds, butterflies, and plant poisons: A study in ecological chemistry. *Zoologica* 49:137–59.

Brower, L. P., J. V. Z. Brower, and C. T. Collins. 1963. Experimental studies of mimicry. 7. Relative palatability and Müllerian mimicry among Neotropical butterflies of the subfamily Heliconiinae. *Zoologica* 48:65–84.

Brower, L. P., J. V. Z. Brower, and F. P. Cranston. 1965. Courtship behavior of the queen butterfly, *Danaus gilippus berenice* (Cramer). *Zoologica* 50:1–39.

Brower, L. P., W. H. Calvert, L. E. Hedrick, and J. Christian. 1977. Biological observations on an overwintering colony of monarch butterflies (*Danaus plexippus*, L., Danaidae) in Mexico. *J. Lep. Soc.* 31:232–42.

Brower, L. P., and S. C. Glazier. 1975. Localization of heart poisons in the monarch butterfly. *Science* 188:19–25.

Brower, L. P., W. N. Ryerson, L. L. Coppinger, and S. C. Glazier. 1968. Ecological chemistry and the palatability system. *Science* 161:1349–51.

Brower, L. P., J. N. Seiber, C. J. Nelson, S. P. Lynch, and P. M. Tuskes. 1982. Plant-determined variation in the cardenolide content, thin-layer chromatography profiles, and emetic potency of monarch butterflies, *Danaus plexippus,* reared on the milkweed, *Asclepias eriocarpa* in California. *J. Chem. Ecol.* 8:579–633.

Brown, D. D. 1981. Gene expression in eucaryotes. *Science* 211:667–74.

Brown, F. M. 1974. An invasion of eastern Colorado by *Vanessa cardui* (Nymphalidae). *J. Lep. Soc.* 28:175.

Brown, K. S., and W. W. Benson. 1974. Adaptive polymorphism associated with Müllerian mimicry in *Heliconius numata* (Lepid. Nymph.). *Biotropica* 6:205–28.

Brussard, P. F., and P. R. Ehrlich. 1970. Adult behavior and population structure in *Erebia epipsodea* (Lepidoptera: Satyridae). *Ecology* 51:880–85.

Brussard, P. F., P. R. Ehrlich, and M. C. Singer. 1974. Adult movements and population structure in *Euphydryas editha*. *Evolution* 28:408–15.

Brussard, P. F., and M. A. Sharp. 1972. Weather and the "regulation" of subalpine populations. *Ecology* 53:243–47.

Burns, J. M. 1972. Intra- and interspecific variations in highly polymorphic esterases of butterflies. *XVII Intern. Congr. Zool.* Theme No. 5, molecular studies of differences between species.

Byers, G. W. 1971. A migration of *Kricogonia castalia* (Pieridae) in northern Mexico. *J. Lep. Soc.* 25:124–25.

Calvert, W. 1974. The external morphology of foretarsal receptors involved with host discrimination by the nymphalid butterfly, *Chlosyne lacinia*. *Annals of E.S.A.* 67:853–56.

Calvert, W. H., L. E. Hedrick, and L. P. Brower. 1979. Mortality of the monarch butterfly (*Danaus plexippus* L.): Avian predation at five overwintering sites in Mexico. *Science* 204:847–51.

Calvert, W. H., W. Zuchowski, and L. Brower. 1982. The impact of forest thinning on microclimate in monarch butterfly (*Danaus plexippus* L.) overwintering areas of Mexico. *Boletin de la sociedad botanica de Mexico* 42:11–18.

Charlesworth, D., and B. Charlesworth. 1975. Theoretical genetics of Batesian mimicry I. Single locus models. *J. Theoret. Biol.* 55:283–303.

Chew, F. S. 1975. Coevolution of pierid butterflies and their cruciferous foodplants. I. The relative quality of available resources. *Oecologia* 20:117–28.

———. 1977. Coevolution of pierid butterflies and their cruciferous foodplants. II. The distribution of eggs on potential foodplants. *Evolution* 31:568–79.

Chun, M. W., and L. M. Schoonhoven. 1973. Tarsal contact chemosensory hairs of the large white butterfly *Pieris brassicae* and their possible role in oviposition behavior. *Ent. Exp. and Appl.* 16:343–57.

Claassens, A. J. M., and C. G. C. Dickson. 1977. A study of the myrmecophilous behavior of the immature stages of *Aloeides thyra* (Lep.: Lycaenidae) with special reference to the function of the retractile biology of the species. *Entomol. Rec. J. Var.* 89:225–31.

Clarke, C. A., and M. Rothschild. 1980. A new mutant of *Danaus plexippus* ssp. *erippus* (Cramer). *J. Lep. Soc.* 34:224–29.

Clarke, C. A., and P. M. Sheppard. 1972. Genetic and environmental factors influencing pupal colour in the swallowtail butterflies *Battus philenor* L. and *Papilio polytes* L. *J. Entomol.* 46:123–33.

Clench, H. K. 1975. Introduction. In *The butterflies of North America*, ed. and ill. W. H. Howe, 1–72. New York: Doubleday and Co.

———. 1976a. Fugitive color in the males of certain pieridae. *J. Lep. Soc.* 30:88–90.

———. 1976b. Nathalis iole (Pieridae) in the southwestern United States and the Bahamas. *J. Lep. Soc.* 30:121–25.

Cody, M. L., and J. M. Diamond. 1975. Introduction. In *Ecology and evolution of communities*, ed. M. L. Cody and J. M. Diamond, 1–14. Cambridge, Mass.: Belknap.

Cohen, J. A., and L. P. Brower. 1982. Oviposition and larval success of wild monarch butterflies (Lepidoptera: Danaidae) in relation to host plant size and cardenolide concentration. *J. Kans. Ent. Soc.* 55:343–48.

Crane, J. 1955. Imaginal behavior of a Trinidad butterfly, *Heliconius erato hydara* Hewitson, with special reference to the social use of color. *Zoologica* 40:167–96.

Craw, R. C. 1975. Lepidoptera feeding at stream margins in New Zealand. *J. Lep. Soc.* 29:198.

Dalton, S. 1975. *Borne on the wind.* New York: Reader's Digest Press.

Davies, T. W., and P. H. Arnaud, Jr. 1967. A remarkable aberrant female of *Speyeria nokomis nokomis* (Edwards) (Lepidoptera: Nymphalidae). *Pan-Pacific Entomologist* 43:177–81.

Dempster, J. P. 1967. The control of *Pieris rapae* with DDT. I. The natural mortality of the young stages of *Pieris*. *J. Appl. Ecol.* 4:485–500.

Dethier, V. G. 1975. The monarch revisited. *J. Kans. Ent. Soc.* 48:129–40.

Dolinger, P. M., P. R. Ehrlich, W. F. Fitch, and D. E. Breedlove. 1973. Alkaloid and predation patterns in Colorado lupine populations. *Oecologia* 13:191–204.

Douglas, M. M. 1978. The behavioral and physiological strategies of thermoregulation in butterflies. Ph.D. diss., University of Kansas.

―――. 1979. Hot butterflies. *Natural History* 88:56–65.

―――. 1981. Thermoregulatory significance of thoracic lobes in the evolution of insect wings. *Science* 211:84–86.

―――. 1983. Defense of bracken fern by arthropods attracted to axillary nectaries. *Psyche* 90:313–20.

Douglas, M. M., and J. W. Grula. 1978. Thermoregulatory adaptations allowing ecological range expansion by the pierid butterfly *Nathalis iole* Boisduval. *Evolution* 32:776–83.

Downes, J. A. 1964. Arctic insects and their environment. *Can. Ent.* 96:279–307.

―――. 1973. Lepidoptera feeding at puddle margins, dung and carrion. *J. Lep. Soc.* 27:89–99.

Downey, J. C. 1957. Observations and new records of hymenopteran parasites of lycaenid eggs (Lepidoptera). *Trans. Ill. State Acad. Sci.* 50:299–300.

―――. 1962a. Host-plant relations on data for butterfly classifications. *Syst. Zool.* 11:150–59.

―――. 1962b. Myrmecophily in *Plebejus (Icaricia) icarioides* (Lepidop.: Lycaenidae). *Entomol. News* 73:57–66.

―――. 1965a. Insect polymorphism. *Proc. N. Cen. Branch E.S.A.* 20:82.

―――. 1965b. Thrips utilize exudations of Lycaenidae. *Entomol. News* 76:25–27.

―――. 1966. Sound production in pupae of Lycaenidae. *J. Lep. Soc.* 20:129–55.

Downey, J. C., and A. C. Allyn. 1975. Wing-scale morphology and nomenclature. *Bull. Allyn Mus.* 31:1–32.

―――. 1981. Eggs of Riodinidae. *J. Lep. Soc.* 34:133–45.

Downey, J. C., and D. B. Dunn. 1964. Variation in the lycaenid butterfly, *Plebejus icarioides*. III. Additional data on food-plant specificity. *Ecology* 45:172–78.

Downey, J. C., and W. C. Fuller. 1961. Variation in *Plebejus icarioides* (Lycaenidae). I. Foodplant specificity. *J. Lep. Soc.* 15:34–42.

Drummond, B. A., III. 1976. Comparative ecology and mimetic relationships of ithomiine butterflies in eastern Ecuador. Ph.D. diss., University of Florida, Gainesville.

Dunlap-Pianka, H. L. 1979. Ovarian dynamics in *Heliconius* butterflies: Correlations among daily oviposition rates, egg weights, and quantitative aspects of oögenesis. *J. Insect Physiol.* 25:741–49.

Dunlap-Pianka, H. L., H. C. Boggs, and L. E. Gilbert. 1977. Ovarian dynamics in Heliconiine butterflies: Programmed senescence versus eternal youth. *Science* 197:487–90.

Durden, C. J., and H. Rose. 1978. Butterflies from the Middle Eocene: The earliest occurrence of fossil Papilionoidea (Lepidoptera). *Pearce-Sellards Series* 29:1–25.

Edgar, J. A., and C. C. J. Culvenor. 1974. Pyrrolizidine ester alkaloids in danaid butterflies. *Nature* 248:614–16.

Ehrlich, P. R. 1961. Intrinsic barriers to dispersal in the checkerspot butterfly. *Science* 134:108–9.

———. 1965. The population biology of the butterfly *Euphydryas editha*. II. The structure of the Jasper Ridge colony. *Evolution* 19:327–36.

———. 1979. The butterflies of Jasper Ridge. *Coevolution Quarterly*, Summer, pp. 50–55.

Ehrlich, P. R., and A. H. Ehrlich. 1961. *How to know the butterflies*. Dubuque, Iowa: Wm. C. Brown Co.

———. 1982. Lizard predation on tropical butterflies. *J. Lep. Soc.* 36:148–52.

Ehrlich, P. R., and L. E. Gilbert. 1973. Population structure and dynamics of the tropical butterfly *Heliconius ethilla*. *Biotropica* 5:69–82.

Ehrlich, P. R., and P. H. Raven. 1965. Butterflies and plants: A study in coevolution. *Evolution* 18:586–608.

———. 1967. Butterflies and plants. *Sci. Amer. New York*. 216:105–13.

———. 1969. Differentiation of populations: Gene flow seems to be less important in speciation than the neo-Darwinians thought. *Science* 165:1228–31.

Ehrlich, P. R., and R. R. White. 1980. Colorado checkerspot butterflies, isolation, neutrality and the biospecies. *Amer. Nat.* 115:328–41.

Ellis, S. L. 1973. Field observations on *Colias alexandra* Edwards (Pieridae). *J. Lep. Soc.* 28:114–24.

Emmel, T. C. 1975. *Butterflies*. New York: Alfred A. Knopf.

Emmel, T. C., and R. A. Wobus. 1966. A southward migration of *Vanessa cardui* in late summer and fall, 1965. *J. Lep. Soc.* 20:123–24.

Evans, W. H. 1975. Seasonal forms of *Anthocaris sara* (Pieridae). *J. Lep. Soc.* 29:52–55.

Faegri, K., and L. van Der Pijl. 1971. *The principles of pollination ecology*, 2d rev. ed. Oxford: Pergamon Press.

Fales, J. H. 1976. More records of butterflies as prey for ambush bugs (Heteroptera). *J. Lep. Soc.* 30:147–49.

Feeny, P. 1975. Biochemical coevolution between plants and their insect herbivores. In *Coevolution of animals and plants*, ed. L. E. Gilbert and P. H. Raven. Austin and London: University of Texas Press.

Ferris, C. D. 1969. Some additional notes in mating behavior in butterflies. *J. Lep. Soc.* 23:271–72.

———. 1974. A note on habitat and geography. *J. Lep. Soc.* 28:166–67.

Ferris, C. D., and F. M. Brown. 1981. *Butterflies of the Rocky Mountain states*. Norman, Oklahoma: University of Oklahoma Press.

Field, W. D., and J. Herrera. 1977. The pierid butterflies of the genera *Hypsochila* Ureta, *Phulia* Herrich-Schaeffer, *Infraphulia* Field, *Pierphulia* Field, and *Piercolias* Staudinger. *Smithsonian Contrib. Zool.* 232:1–64.

Fink, L. S., and L. P. Brower. 1981. Birds can overcome the cardenolide defence of monarch butterflies in Mexico. *Nature* 291:67–70.

Fox, R. M. 1966. Forelegs of butterflies. I. Introduction: Chemoreception. *J. Res. Lepid.* 5:1–12.

Funk, R. S. 1968. Overwintering of monarch butterflies in southwestern Arizona. *J. Lep. Soc.* 22:63–64.

Futuyama, D. J. 1976. Food plant specialization and environmental predictability in Lepidoptera. *Amer. Nat.* 110:285–92.

Gadgil, M. 1972. The function of communal roosts: Relevance of mixed roosts. *Ibis* 114:531–33.

Gilbert, L. E. 1971. Butterfly-plant coevolution: Has *Passiflora adenopoda* won the selectional race with heliconiine butterflies? *Science* 172:585–86.

———. 1972. Pollen feeding and reproductive biology of *Heliconius* butterflies. *Proc. Natl. Acad. Sci. U.S.A.* 69:1403–7.

———. 1975. Ecological consequences of a coevolved mutualism between butterflies and plants. In *Coevolution of animals and plants*, ed. L. E. Gilbert and P. H. Raven, 210–40. Austin and London: University of Texas Press.

———. 1976. Postmating female odor in *Heliconius* butterflies: A male-contributed antiaphrodisiac? *Science* 193:419–20.

———. 1979. Developments of theory in the analysis of insect-plant interaction. In *Analysis of ecological systems*, ed. D. Horn, R. Mitchell, and G. Stairs, 117–54. Columbus: Ohio State University Press.

———. 1980. Food web organization and the conservation of Neotropical diversity. In *Conservation biology: An evolutionary-ecological perspective*, ed. M. E. Soule and B. A. Wilcox, 11–33. Sunderland: Sinauer.

———. 1982. The coevolution of a butterfly and a vine. *Scientific American* 247:110–21.

———. 1984. The biology of butterfly communities. Royal Entomological Society Symposium. In *Biology of butterflies*, ed. R. I. Vane-Wright and P. R. Ackery. Royal Entomological Society Symposium. London: Academic Press.

Gilbert, L. E., and M. C. Singer. 1973. Dispersal and gene flow in a butterfly species. *Amer. Nat.* 107:58–72.

———. 1975. Butterfly ecology. *Ann. Rev. Ecol. and Syst.* 6:365–97.

Gilbert, L. E., and J. T. Smiley. 1978. Determinants of local diversity in phytophagous insects: Host specialists in tropical environments. In *Diversity of insect faunas*, ed. L. A. Mound and N. Waloff, 89–104. Symposia of the Royal Entomological Society of London, No. 9. Oxford: Blackwell.

Gilbert, L. I. 1964. Physiology of growth and development: Endocrine aspects. In *The physiology of insecta*, vol. I, ed. M. Rockstein. New York: Academic Press.

Goldschmidt, R. 1927. *Physiologische Theorie der Vererbung*. Berlin: Springer-Verlag.

Grula, J. W. 1978. The inheritance of traits maintaining ethological isolation between two species of *Colias* butterflies. Ph.D. diss., University of Kansas.

Grula, J. W., J. D. McChesney, and O. R. Taylor, Jr. 1980. Aphrodisiac pheromones of the sulfur butterflies *Colias eurytheme* and *C. philodice* (Lepidoptera, Pieridae). *J. Chem. Ecol.* 6:241–56.

Grula, J. W., and O. R. Taylor, Jr. 1979. The inheritance of pheromone production in the sulfur butterflies *Colias eurytheme* and *C. philodice*. *Heredity* 42:359–71.

Grula, J. W., and O. R. Taylor. 1980a. A micromorphological and experimental study of the antennae of the sulfur butterflies, *Colias eurytheme* and *C. philodice* (Lepidoptera, Pieridae). *J. Kans. Ent. Soc.* 53:476–84.

————. 1980b. The effect of X-chromosome inheritance on mate-selection behavior in the sulfur butterflies, *Colias eurytheme* and *C. philodice*. *Evolution* 34:688–95.

————. 1980c. Some characteristics of hybrids derived from the sulfur butterflies, *Colias eurytheme* and *C. philodice:* Phenotypic effects on the X-chromosome. *Evolution* 34:673–87.

Haber, W. A. 1978. Evolutionary ecology of tropical mimetic butterflies (Lepidoptera: Ithomiinae). Ph.D. diss., University of Minnesota.

Hafernik, J. E., Jr. 1983. Phenetics and ecology of hybridization in buckeye butterflies (Lepidoptera: Nymphalidae). *Univ. Calif. Pubs. in Ent.* 96:1–109.

Harcourt, D. G. 1966. Major factors in survival of the immature stages of *Pieris rapae* (L.). *Can. Ent.* 98:653–62.

Harvey, D. J., and T. A. Webb. 1980. Ants associated with *Harkenclenus titus, Glaucopsyche lygdamus,* and *Celestrina argiolus* (Lycaenidae). *J. Lep. Soc.* 34:371–72.

Hayes, J. L. 1979. A synthesis of diapause strategies in Lepidoptera. *Kans. Acad. Sci.* 82:93 (abstract).

————. 1980. Some aspects of the biology of the developmental stages of *Colias alexandra* (Pieridae). *J. Lep. Soc.* 34:345–52.

————. 1981. The population ecology of a natural population of the pierid butterfly, *Colias alexandra*. *Oecologia* 49:188–200.

————. 1982a. Diapause and diapause dynamics of *Colias alexandra* (Lepidoptera: Pieridae). *Oecologia* 53:317–22.

————. 1982b. A study of the relationships of diapause phenomena and other life history characters in temperate butterflies. *Amer. Nat.* 120:160–70.

Hazel, W. N. 1977. The genetic basis of pupal colour dimorphism and its maintenance by natural selection in *Papilio polyxenes* (Papilionidae: Lepidoptera). *Heredity* 38:227–36.

Heinrich, B. 1971a. Temperature regulation of the sphinx moth, *Manduca sexta*. I. Flight energetics and body temperature during free and tethered flight. *J. Exp. Biol.* 54:141–52.

————. 1971b. Temperature regulation of the sphinx moth, *Manduca sexta*. II. Regulation of heat loss by control of blood circulation. *J. Exp. Biol.* 54:153–66.

————. 1979. *Bumblebee economics*. Cambridge: Harvard University Press.

Hiam, A. W. 1982. Airborne models and flying mimics. *Nat. Hist.* 91:42–49.

Hinton, H. E. 1976. Possible significance of the red patches of the female crab spider, *Misumena vatia*. *J. Zool.* 180:35–39.

Hoffmann, R. J. 1973. Environmental control of seasonal variation in the butterfly *Colias eurytheme*. I. Adaptive aspects of a photoperiod response. *Evolution* 27:387–97.

Holland, W. J. 1899. *The butterfly book*. New York: Doubleday and McClure.

Hong, J. W., and A. P. Platt. 1975. Critical photoperiod and day-length threshold differences between northern and southern populations of the butterfly, *Limenitis archippus*. *J. Insect Physiol.* 21:1159–65.

Howe, W. H. 1964. Migration of *Kricogonia lyside* in Mexico (Pieridae). *J. Lep. Soc.* 18:26–27.

———. 1965. Status of *Agraulis vanillae* in Missouri and Kansas. *J. Lep. Soc.* 19:33–34.

———. 1975. *The butterflies of North America.* New York: Doubleday and Co.

Ivie, G. W., D. L. Bull, R. C. Beier, N. W. Pryor, and E. H. Oertli. 1983. Metabolic detoxification: Mechanism of insect resistance to plant psoralens. *Science* 221:374–76.

Janzen, D. H. 1983. *Costa Rican natural history.* Chicago: University of Chicago Press.

Jennings, D. T., and M. E. Toliver. 1976. Crab spider preys on *Neophasia menapia* (Pieridae). *J. Lep. Soc.* 30:236–37.

Jobe, J. B. 1977. On "honeydew-panting" in Lepidoptera. *Entomol. Gaz.* 28:8.

Johnson, F. M., and J. M. Burns. 1966. Electrophoretic variation in esterase of *Colias eurytheme* (Pieridae). *J. Lep. Soc.* 20:207–11.

Kammer, A. E. 1970. Thoracic temperature, shivering, and flight in the monarch butterfly, *Danaus plexippus* (L.). *Z. Vergl. Physiologie* 68:334–44.

———. 1971. Influence of acclimation temperature on the shivering behavior of the butterfly *Danaus plexippus* (L.). *Z. Vergl. Physiologie* 72:364–69.

Kane, S. 1982. Notes on the acoustic signals of a Neotropical satyrid butterfly. *J. Lep. Soc.* 36:200–206.

Kanz, J. E. 1977. The orientation of migrant and non-migrant monarch butterflies, *Danaus plexippus* (L.). *Psyche* 84:120–41.

Kevan, P. G., and J. D. Shorthouse. 1970. Behavioral thermoregulation by high arctic butterflies. *Arctic* 23:268–79.

Kingsolver, J. G. 1983a. Ecological significance of flight activity in *Colias* butterflies: Implications for reproductive strategy and population structure. *Ecology* 64:546–51.

———. 1983b. Thermoregulation and flight in *Colias* butterflies: Elevational patterns and mechanistic limitations. *Ecology* 64:534–45.

Kingsolver, J. G., and T. L. Daniel. 1979. On the mechanics and energetics of nectar feeding in butterflies. *J. Theoret. Biol.* 76:167–79.

Kingsolver, J. G., and M. A. R. Koehl. 1985. Aerodynamics, thermoregulation, and the evolution of insect wings: Differential scaling and evolutionary change. *Evolution* 39:488–504.

Kingsolver, J. G., and W. B. Watt. 1983. Thermoregulatory strategies in *Colias* butterflies: Thermal stress and the limits to adaptation in temporally varying environments. *Amer. Nat.* 121:32–55.

———. 1984. Mechanistic restraints and optimality models: Thermoregulatory strategies in *Colias* butterflies. *Ecology* 65:1835–39.

Klots, A. B. 1951. *A field guide to the butterflies of North America, east of the Great Plains.* Boston: Houghton Mifflin Co.

Krivda, W. V. 1976. A migration of *Vanessa cardui* (Nymphalidae). *J. Lep. Soc.* 30:312.

Kukalová-Peck, J. 1978. Origin and evolution of insect wings and their relation to metamorphosis, as documented by the fossil record. *J. Morph.* 156:53–125.

Labine, P. A. 1968. The population biology of the butterfly, *Euphydryas editha.* VIII. Oviposition and its relation to patterns of oviposition in other butterflies. *Evolution* 22:799–805.

Larsen, T. B. 1982. False head butterflies: The case of *Oxylides faunas* Drury (Lycaenidae). *J. Lep. Soc.* 36:238–39.

Levin, D. A., and D. E. Berube. 1972. *Phlox* and *Colias:* The efficiency of a pollination system. *Evolution* 26:242–50.

Levins, R., and R. MacArthur. 1969. An hypothesis to explain the incidence of monophagy. *Ecology* 50:910–22.

Lewin, R. 1982. Can genes jump between eukaryotic species? *Science* 217:42–43.

Lewontin, R. C., and L. C. Birch. 1966. Hybridization as a source of variation for adaptation to new environments. *Evolution* 20:315–36.

MacArthur, R., and E. O. Wilson. 1967. *The theory of island biogeography.* Princeton, N.J.: Princeton University Press.

Mackay, D. A. 1985. Prealighting search behavior and host plant selection by ovipositing *Euphydryas editha* butterflies. *Ecology* 66:142–51.

Maeki, K., and C. L. Remington. 1960. Studies of the chromosomes of North American Rhopalocera. 2. Hesperiidae, Megathymidae, and Pieridae. *J. Lep. Soc.* 14:37–57.

Malicky, H. 1970. New aspects on the association between lycaenid larvae (Lycaenidae) and ants (Formicidae, Hymenoptera). *J. Lep. Soc.* 24:190–202.

Manley, T. R. 1971. Dragonfly attacks *Limenitis* defending its territory. *J. Lep. Soc.* 25:146–47.

Masters, J. H. 1969. Season variation of *Colias cesonia therapis* in Venezuela (Pieridae). *J. Lep. Soc.* 23:251–53.

———. 1979. A documentation of biennialism in *Boloria polaris* (Nymphalidae). *J. Lep. Soc.* 33:167–69.

Mayr, E. 1963. *Animal species and evolution.* Cambridge: Harvard University Press.

Miller, L. D., and H. K. Clench. 1968. Some aspects of mating behavior in butterflies. *J. Lep. Soc.* 22:125–32.

Monroe, E. G. 1948. The geographical distribution of butterflies in the West Indies. Ph.D. diss., Cornell University.

Murphy, D. D. 1984. Butterflies and their nectar plants: The role of the checkerspot butterfly *Euphydryas editha* as a pollen vector. *Oikos* 43:113–17.

Murphy, D. D., and R. R. White. 1986. Rainfall, resources, and dispersal in southern populations of *Euphydryas editha* (Lepidoptera: Nymphalidae). *Pan-Pacific Entomologist* (in press).

Myers, J. 1968. The structure of antennae of the Florida Queen butterfly, *Danaus gilippus berenice. J. Morph.* 125:315–28.

———. 1969. Distribution of foodplant chemoreceptors on the female Florida Queen butterfly, *Danaus gilippus berenice* (Nymphalidae). *J. Lep. Soc.* 23:196–98.

Myers, J., and L. P. Brower. 1969. A behavioral analysis of the courtship pheromone receptors of the queen butterfly, *Danaus gilippus berenice. J. Insect Physiol.* 15:2117–30.

Nakamura, I. 1976. Female anal hair tuft in *Nordmannia myrtale* (Lycaenidae): Egg-camouflaging function and taxonomic significance. *J. Lep. Soc.* 30:305–9.

Nardi, J. B., and F. C. Kafatos. 1976. Polarity and gradients in lepidopteran wing epidermis. I. Changes in graft polarity, form, and cell density accompanying transpositions and reorientations. *J. Embryol. Exp. Morphol.* 36:469–87.

Neck, R. W. 1976a. Larval morph variation in *Chlosyne lacinia* (Nymphalidae). *J. Lep. Soc.* 30:91–94.

———. 1976b. Summer monarch (*Danaus plexippus*) in southern Texas (Danaidae). *J. Lep. Soc.* 30:137.

———. 1976c. Nocturnal activity of a monarch butterfly (Danaidae). *J. Lep. Soc.* 30:235–36.

———. 1978. Additional function of the lepidopteran proboscis. *J. Lep. Soc.* 32:310–11.

———. 1980. Utilization of grass inflorescences as adult resources by Rhopalocera. *J. Lep. Soc.* 34:261–62.

———. 1983. Significance of visits by hackberry butterflies (Nymphalidae: *Asterocampa*) to flowers. *J. Lep. Soc.* 37:269–74.

Neck, R. W., G. L. Bush, and B. A. Drummond. 1971. Epistasis, associated lethals and brood effect in larval color polymorphism of the patch butterfly, *Chlosyne lacinia. Heredity* 26:73–84.

Nelson, C. J., J. N. Seiber, and L. P. Brower. 1981. Seasonal and intraplant variation of cardenolide content in the California milkweed, *Asclepias eriocarpa*, and implications for plant defense. *J. Chem. Ecol.* 7:981–1010.

Nijhout, H. F. 1978. Wing pattern formation in Lepidoptera: A model. *J. Exp. Zool.* 206:119–36.

———. 1980a. Pattern formation on lepidopteran wings: Determination of an eyespot. *Develop. Biol.* 80:267–74.

———. 1980b. Ontogeny of the color pattern on the wings of *Precis coenia* (Lepidoptera: Nymphalidae). *Develop. Biol.* 80:275–88.

———. 1981. The color patterns of butterflies and moths. *Scientific American* 245:140–51.

Novak, I. 1980. *A field guide in colour to butterflies and moths.* London: Octopus Books Ltd.

Ohsaki, N. 1979. Comparative population studies of three *Pieris* butterflies, *P. rapae, P. melete* and *P. napi*, living in the same area. I. Ecological requirements for habitat resources in the adults. *Res. Popul. Ecol.* 20:278–96.

Oliver, C. G. 1972. Genetic and phenotypic differentiation and geographic distance in four species of Lepidoptera. *Evolution* 26:221–41.

———. 1976. Photoperiodic regulation of seasonal polyphenism in *Phyciodes tharos* (Nymphalidae). *J. Lep. Soc.* 30:260–63.

———. 1979. Genetic differentiation and hybrid viability within and between some lepidopteran species. *Amer. Nat.* 114:681–94.

Opler, P. A., and G. O. Krizek. 1984. *Butterflies east of the Great Plains.* Baltimore and London: Johns Hopkins University Press.

Pan, M. L., and G. R. Wyatt. 1971. Juvenile hormone induces vitellogenin synthesis in the monarch butterfly. *Science* 174:503–5.

Papageorgis, C. 1975. Mimicry in Neotropical butterflies. *Amer. Scientist* 63:522–32.

Pierce, N. E., and P. S. Mead. 1981. Parasitoids as selective agents in the symbiosis between lycaenid butterfly larvae and ants. *Science* 211:1185–87.

Platt, A. P. 1975. Monomorphic mimicry in Nearctic *Limenitis* butterflies: Experimental hybridization of the *L. arthemis-astyanax* complex with *L. archippus. Evolution* 29:120–41.

Pliske, T. E., and M. M. Salpeter. 1971. The structure and development of the hair-pencil glands in males of the queen butterfly, *Danaus gilippus berenice*. *J. Morph.* 134:215–42.

Price, P. W., C. E. Bouton, P. Gross, A. McPherson, J. N. Thompson, and A. E. Weiss. 1980. Interactions among three trophic levels: Influence of plants on interaction between insect herbivores and natural enemies. *Ann. Rev. Ecol. Syst.* 11:41–65.

Proctor, N. S. 1976. Mass hibernation site for *Nymphalis vau-album* (Nymphalidae). *J. Lep. Soc.* 30:126.

Pyle, R. M. 1973. *Boloria selene* (Nymphalidae) ambushed by a true bug (Heteroptera). *J. Lep. Soc.* 27:305–7.

———. 1981. *The Audubon Society field guide to North American butterflies*, Chanticleer Press edition. New York: Alfred A. Knopf.

Ratcliffe, D. 1979. The end of the large blue butterfly. *New Sci.* 84:457–58.

Rathcke, B. J., and R. W. Poole. 1975. Coevolutionary race continues: Butterfly larval adaptations to plant trichomes. *Science* 187:175–76.

Rausher, M. D. 1978. Search image for leaf shape in a butterfly. *Science* 200:1071–73.

———. 1979. Egg recognition: Its advantage to a butterfly. *Animal Behaviour* 27:1034–40.

———. 1980. Host abundance, juvenile survival, and oviposition preference in *Battus philenor*. *Evolution* 34:342–55.

———. 1981. Host plant selection by *Battus philenor:* The roles of predation, nutrition, and plant chemistry. *Ecological Monographs* 51:1–20.

———. 1982. Population differentiation in *Euphydryas editha* butterflies: Larval adaptation to different hosts. *Evolution* 36:581–90.

———. 1983. Alteration of oviposition behavior by *Battus philenor* butterflies in response to variation in host-plant density. *Ecology* 64:1028–34.

Rausher, M. D., D. A. Mackay, and M. C. Singer. 1981. Pre- and post-alighting host discrimination by *Euphydryas editha* butterflies: The behavioural mechanisms causing clumped distributions of egg clusters. *Animal Behaviour* 29:1220–28.

Rausher, M. D., and D. R. Papaj. 1983a. Host plant selection by *Battus philenor* butterflies: Evidence for individual differences in foraging behaviour. *Animal Behaviour* 31:341–47.

———. 1983b. Demographic consequences of conspecific host discrimination by *Battus philenor* butterflies. *Ecology* 64:1402–10.

Ray, T. S. and C. C. Andrews. 1980. Antbutterflies: Butterflies that follow army ants to feed on antbird droppings. *Science* 210:1147–48.

Reinthal, W. J. 1963. About the "pumping action" of a *Papilio* at water. *J. Lep. Soc.* 17:35.

Remington, C. L. 1973. Ultraviolet reflectance in mimicry and sexual signals in the Lepidoptera. *J. N. Y. Ent. Soc.* 81:124.

Richards, O. W. 1940. The biology of the small white butterfly (*Pieris rapae*), with special reference to the factors controlling its abundance. *J. Anim. Ecol.* 9:243–88.

Riley, T. J. 1980. Effects of long and short day photoperiods on the seasonal dimorphism of *Anaea andria* (Nymphalidae) from central Missouri. *J. Lep. Soc.* 34:330–37.

Robbins, R. K. 1980. The lycaenid "false head" hypothesis: Historical review and quantitative analysis. *J. Lep. Soc.* 34:194–208.

———. 1981. The "false head" hypothesis: Predation and wing pattern variation of lycaenid butterflies. *Amer. Nat.* 118:770–75.

Rodman, J. E., L. P. Brower, and J. Frey. 1982. Cardenolides in North American *Erysimum* (Cruciferae), a preliminary chemotaxonomic report. *Taxon* 31:507–16.

Rodman, J. E., and F. S. Chew. 1980. Phytochemical correlates of herbivory in a community of native and naturalized Cruciferae. *Biochem. Syst. and Ecol.* 8:43–50.

Ross, G. N. 1963. Evidence for lack of territoriality in two species of *Hamadryas* (Nymphalidae). *J. Res. Lepid.* 2:241–46.

Rothschild, M., and L. M. Schoonhoven. 1977. Assessment of egg load by *Pieris brassicae* (Lepidoptera: Pieridae). *Nature* 266:352–55.

Rutowski, R. L. 1979. Courtship behavior of the checkered white, *Pieris protodice* (Pieridae). *J. Lep. Soc.* 33:42–49.

———. 1980. Male scent-producing structures in *Colias* butterflies: Function, localization and adaptive features. *J. Chem. Ecol.* 6:13–26.

Schwanwitsch, B. N. 1924. On the ground-plan of wing-pattern in nymphalids and certain other families of rhopalocerous Lepidoptera. *Proc. Zoo. Soc. London* 34:509–28.

Scott, J. 1973. Down-valley flight of adult theclini (Lycaenidae) in search of nourishment. *J. Lep. Soc.* 27:283–87.

Scott, J. W., and P. A. Opler. 1975. Population biology and adult behavior of *Lycaena xanthoides* (Lycaenidae). *J. Lep. Soc.* 29:63–66.

Scriber, J. M., and P. Feeny. 1979. Growth of herbivorous caterpillars in relation to feeding specialization and to growth form of their food plants. *Ecology* 60:829–50.

Scudder, S. H. 1889. *The butterflies of the eastern United States and Canada with special reference to New England.* 3 vols. Cambridge, Mass.: privately published.

Sevastopulo, D. G. 1974. Lepidoptera feeding at puddle-margins, dung and carrion. *J. Lep. Soc.* 28:167–68.

Shapiro, A. M. 1971. Occurrence of a latent polyphenism in *Pieris virginiensis* (Lepidoptera: Pieridae). *Entomol. News.* 82:13–16.

———. 1973. Photoperiodic control of seasonal polyphenism in *Pieris occidentalis* Reakirt (Lepidoptera: Pieridae). *Wasmann J. Biol.* 31:291–99.

———. 1975a. Ecological and behavioral aspects of coexistence in six crucifer-feeding pierid butterflies in the central Sierra Nevada. *Am. Midl. Nat.* 93:424–33.

———. 1975b. The temporal component of butterfly species diversity. In *Ecology and evolution of communities,* ed. M. L. Cody and J. M. Diamond, 181–95. Cambridge, Mass.: Harvard University Press.

————. 1975c. Developmental and phenotypic responses to photoperiod in uni- and bivoltine *Pieris napi* (Lepidoptera: Pieridae) in California. *Trans. R. Ent. Soc. London* 127:65–71.

————. 1976a. Seasonal polyphenism. *Evol. Biol.* 9:259–333.

————. 1976b. The biological status of Nearctic taxa in the *Pieris protodice-occidentalis* group (Pieridae). *J. Lep. Soc.* 30:289–300.

————. 1977. Phenotypic induction of *Pieris napi* L.: Role of temperature and photoperiod in a coastal California population. *Ecol. Ent.* 2:219–24.

————. 1978a. The evolutionary significance of redundancy and variability in phenotypic-induction mechanisms of pierid butterflies (Lepidoptera). *Psyche* 85:275–83.

————. 1978b. Weather and the lability of breeding populations of the checkered white butterfly, *Pieris protodice* Boisduval and LeConte. *J. Res. Lepid.* 17:1–23.

————. 1980a. Genetic incompatibility between *Pieris callidice* and *Pieris occidentalis nelsoni:* Differentiation within a periglacial relict complex (Lepidoptera: Pieridae). *Can. Ent.* 112:463–68.

————. 1980b. Convergence in pierine polyphenisms (Lepidoptera). *J. Nat. Hist.* 14:781–802.

————. 1980c. Physiological and developmental responses to photoperiod and temperature as data in phylogenetic and biogeographic inference. *Syst. Zool.* 29:335–41.

————. 1980d. Egg load assessment and carryover diapause in *Anthocharis* (Pieridae). *J. Lep. Soc.* 34:307–15.

————. 1981a. Egg-mimics of *Streptanthus* (Cruciferae) deter oviposition by *Pieris sisymbrii* (Lepidoptera: Pieridae). *Oecologia* 48:142–43.

————. 1981b. The pierid red-egg syndrome. *Amer. Nat.* 117:276–94.

————. 1981c. Phenotypic plasticity in temperate and subarctic *Nymphalis antiopa* (Nymphalidae): Evidence for adaptive canalization. *J. Lep. Soc.* 35:124–31.

————. 1982. Redundancy in pierid polyphenisms: Pupal chilling induces vernal phenotype in *Pieris occidentalis* (Pieridae). *J. Lep. Soc.* 36:174–77.

————. 1983. Testing visual species recognition in *Precis* (Lepidoptera: Nymphalidae) using a cold-shock phenocopy. *Psyche* 89:59–65.

Sherman, P., and W. B. Watt. 1973. The thermal ecology of some *Colias* butterfly larvae. *J. Comp. Physiol.* 83:25–40.

Shields, O. 1967. Hilltopping. *J. Res. Lepid.* 6:71–78.

————. 1972. Flower visitation records for butterflies. *Pan-Pacific Entomologist* 48:189–203.

————. 1976. Fossil butterflies and the evolution of Lepidoptera. *J. Res. Lepid.* 15:132–43.

Showalter, A. H., and B. M. Drees. 1980. Bilateral gynandromorphic *Speyeria diana* (Nymphalidae). *J. Lep. Soc.* 34:340–44.

Sibatani, A. 1973. Taxonomic significance of reflective patterns in the compound eye of live butterflies: A synthesis of observations made on species from Japan, Taiwan, Papua New Guinea and Australia. *J. Lep. Soc.* 27:161–75.

Silberglied, R. E. 1977. Communication in the Lepidoptera. In *How animals communicate*, ed. T. A. Sebeok, 362–402. Bloomington: Indiana University Press.

———. 1979. Communication in the ultraviolet. *Ann. Rev. Ecol. Syst.* 10:373–98.

———. 1984. Visual communications and sexual selection among butterflies. In *Biology of butterflies*, ed. R. I. Vane-Wright and P. R. Ackery. Royal Entomological Society Symposium. London: Academic Press.

Silberglied, R. E., A. Aiello, and G. Lamas. 1980. Neotropical butterflies of the genus *Anartia:* Systematics, life histories, and general biology (Lepidoptera: Nymphalidae). *Psyche* 86:219–60.

Silberglied, R. E., A. Aiello, and D. M. Windsor. 1980. Disruptive coloration in butterflies: Lack of support in *Anartia fatima. Science* 209:617–19.

Silberglied, R. E., and O. R. Taylor, Jr. 1973. Ultraviolet differences between the sulphur butterflies, *Colias eurytheme* and *C. philodice*, and a possible isolating mechanism. *Nature* 241:406–8.

———. 1978. Ultraviolet reflection and its behavioral role in the courtship of the sulfur butterflies, *Colias eurytheme* and *C. philodice* (Lepidoptera, Pieridae). *Behav. Ecol. Sociobiol.* 3:203–43.

Sims, S. R. 1980. Diapause dynamics and host plant suitability of *Papilio zelicaon* (Lepidoptera: Papilionidae). *Am. Midl. Nat.* 103:375–84.

Sims, S. R., and A. M. Shapiro. 1983. Seasonal phenology of *Battus philenor* (L.) (Papilionidae) in California. *J. Lep. Soc.* 37:281–88.

Singer, M. C. 1971. Evolution of food-plant preference in the butterfly *Euphydryas editha. Evolution* 25:383–89.

———. 1972. Complex components of habitat suitability within a butterfly colony. *Science* 176:75–77.

Slansky, F., Jr. 1974. Relationship of larval foodplant and voltinism patterns in temperate butterflies. *Psyche* 81:243–53.

Slansky, F. 1976. Phagism relationships among butterflies. *J. N. Y. Ent. Soc.* 84:91–105.

Slansky, F., and P. Feeny. 1977. Stabilization of the role of nitrogen accumulation by larvae of the cabbage butterfly on wild and cultivated food plants. *Ecological Monographs* 47:209–28.

Smart, P. 1975. *The international butterfly book.* New York: Thomas Y. Crowell Co.

Smiley, J. T. 1978a. Plant chemistry and the evolution of host specificity: New evidence from *Heliconius* and *Passiflora. Science* 201:745–47.

———. 1978b. Host plant ecology of *Heliconius* butterflies in northeastern Costa Rica. Ph.D. diss., University of Texas, Austin.

———. 1985. *Heliconius* caterpillar mortality during establishment on plants with and without attending ants. *Ecology* 66:845–949.

Smith, D. A. S. 1975. Sexual selection in a wild population of the butterfly *Danaus chrysippus* L. *Science* 187:664–65.

Snodgrass, R. E. 1935. *Principles of insect morphology.* New York: McGraw-Hill.

Stamp, N. 1977. Aggregation behavior of *Chlosyne lacinia* larvae (Nymphalidae). *J. Lep. Soc.* 31:35–40.

Stamp, N. E. 1984. Interactions of parasitoids and checkerspot caterpillars *Euphydryas* spp. (Nymphalidae). *J. Res. Lepid.* 23:2–18.

Stanton, M. L. 1984. Short-term learning and the searching accuracy of egg-laying butterflies. *Animal Behaviour* 32:33–40.

Süffert, F. 1927. Zur vergleichenden Analyse der Schmetterlingszeichnung. *Biol. Zentr.* 47:385–413.

Swihart, S. L. 1967. Hearing in butterflies (Nymphalidae: *Heliconius, Ageronia*). *J. Insect Physiol.* 13:469–76.

――――. 1972. The neural basis of colour vision in the butterfly, *Heliconius erato. J. Insect Physiol.* 18:1015–25.

Taylor, O. R. 1968. Reproductive diapause in tropical butterflies: An adaptation to a severe dry season. Seminar, University of Connecticut.

――――. 1972. Random vs. non-random mating in the sulfur butterflies, *Colias eurytheme* and *Colias philodice* (Lepidoptera: Pieridae). *Evolution* 26:344–56.

――――. 1973a. A non-genetic "polymorphism" in *Anartia fatima* (Lepidoptera: Nymphalidae). *Evolution* 27:161–64.

――――. 1973b. Reproductive isolation in *Colias eurytheme* and *C. philodice* (Lepidoptera: Pieridae): Use of olfaction in mate selection. *Annals of E.S.A.* 66:621–26.

Tindale, N. B. 1980. Origin of the Lepidoptera, with description of a new Mid-Triassic species and notes on the origin of the butterfly stem. *J. Lep. Soc.* 34:263–85.

――――. 1981. The origin of the Lepidoptera relative to Australia. In *Ecological biogeography of Australia*, ed. A. Keast, 957–75. Dr. W. Junk by Publishers.

Treat, A. E. 1975. *Mites of moths and butterflies*. Ithaca and London: Cornell University Press.

Turner, J. R. 1971. The genetics of some polymorphic forms of the butterflies *Heliconius melpomene* and *H. erato*. 2. The hybridization of subspecies from Surinam and Trinidad. *Zoologica* 56:125–58.

――――. 1975. Communal roosting in relation to warning coloration in two heliconiine butterflies (Nymphalidae). *J. Lep. Soc.* 29:221–26.

――――. 1976. Müllerian mimicry: Classical "beanbag" evolution and the role of ecological islands in adaptive race formation. In *Population genetics and ecology*, ed. S. Karlin and E. Nevo, 185–218. New York and London: Academic Press.

Tyler, H. A. 1975. *The swallowtail butterflies of North America*. Healdsburg, Calif.: Naturegraph Publishers.

Urquhart, F. A., and N. R. Urquhart. 1976a. The overwintering site of the eastern population of the monarch butterfly (*Danaus p. plexippus* L., Danaidae) in southern Mexico. *J. Lep. Soc.* 30:153–58.

――――. 1976b. Migration of butterflies along the gulf coast of northern Florida. *J. Lep. Soc.* 30:59–61.

――――. 1976c. A study of the peninsular Florida populations of the monarch butterfly (*Danaus p. plexippus:* Danaidae). *J. Lep. Soc.* 30:73–88.

Vetter, R. S., and R. L. Rutowski. 1978. External sex brand morphology of three sulphur butterflies (Lepidoptera: Pieridae). *Psyche* 85:383–93.

Waller, D. A., and L. E. Gilbert. 1982. Roost requirement and resource utilization: Observations on a *Heliconius charitonia* L. roost in Mexico (Nymphalidae). *J. Lep. Soc.* 36:178–84.

Wasserthal, L. T. 1974. Heartbeat reversal in insects and the development of heart-rhythm in the adult. *Verh. Dtsch. Zool. Ges.* 67:95–99.

———. 1975. The role of butterfly wings in regulation of body temperature. *J. Insect Physiol.* 21:1921–30.

———. 1976. Heartbeat reversal and its coordination with accessory pulsatile organs and abdominal movements in Lepidoptera. *Experimentia* 32:577–78.

———. 1980. Oscillating haemolymph 'circulation' in the butterfly *Papilio machaon* L. revealed by contact thermography and photocell measurements. *J. Comp. Physiol.* 139:145–63.

Watt, W. B. 1968. Adaptive significance of pigment polymorphism in *Colias* butterflies. I. Variation of melanin pigment in relation to thermoregulation. *Evolution* 22:437–58.

———. 1969. Adaptive significance of pigment polymorphism in *Colias* butterflies. II. Thermoregulation and photoperiodically controlled melanin variation in *Colias eurytheme*. *Proc. Natl. Acad. Sci. U.S.A.* 63:767–74.

Watt, W., D. Han, and B. Tabashnik. 1979. Population structure of pierid butterflies. II. A "native" population of *Colias philodice eriphyle* in Colorado. *Oecologia* 44:44–52.

Watt, W. B., P. C. Hoch, and S. Mills. 1974. Nectar resource use by *Colias* butterflies: Chemical and visual aspects. *Oecologia* 14:353–74.

Weins, J. A. 1976. Population responses to patchy environments. *Ann. Rev. Ecol. Syst.* 7:81–120.

West, D. A., W. M. Snellings, and T. A. Herbeck. 1972. Pupal color dimorphism and its environmental control in *Papilio polyxenes asterius* Stoll (Lepidoptera: Papilionidae). *J. N. Y. Ent. Soc.* 80:205–11.

White, R. R. 1973. Community relationships of the butterfly *Euphydryas editha*. Ph.D. diss., Stanford University.

———. 1980. Inter-peak dispersal in alpine checkerspot butterflies (Nymphalidae). *J. Lep. Soc.* 34:353–62.

White, R. R., and M. P. Levin. 1981. Temporal variation in insect vagility: Implications for evolutionary studies. *Am. Midl. Nat.* 105:348–57.

Wigglesworth, V. B. 1963. The origin of flight in insects. *Proc. R. Ent. Soc. London* 28:23–32.

———. 1972. *The principles of insect physiology*, 7th ed. London: Chapman and Hall.

Wiklund, C. 1972. Pupal coloration in *Papilio machaon* in response to the wavelength of light. *Die Naturwissenschaften*, 1–2.

———. 1975a. Pupal colour polymorphism in *Papilio machaon* L. and the survival in the field of cryptic vs. non-cryptic pupae. *Trans. R. Ent. Soc. London* 127:73–84.

———. 1975b. The evolutionary relationship between adult oviposition preferences and larval host range in *Papilio machaon* L. *Oecologia* 18:185–97.

Wiklund, C., T. Erikson, and H. Lundberg. 1979. The wood white butterfly, *Leptidea sinapsis*, and its nectar plants: A case of mutualism or parasitism? *Oikos* 33:358–62.

Williams, C. B. 1930. *The migration of butterflies.* Edinburgh: Oliver and Boyd.

———. 1970. The migration of the painted lady butterfly, *Vanessa cardui* (Nymphalidae), with special reference to North America. *J. Lep. Soc.* 24:157–75.

Williams, K. S., and L. E. Gilbert. 1981. Insects as selective agents on plant vegetative morphology: Egg mimicry reduces egg laying by butterflies. *Science* 212:467–69.

Yagi, N., and N. Koyama. 1963. *The compound eye of Lepidoptera: Approach from organic evolution.* Tokyo: Shinkyo.

Young, A. M. 1980. Notes on the behavioral ecology of *Perrhybris lypera* (Pieridae) in northeastern Costa Rica. *J. Lep. Soc.* 34:36–47.

Young, A. M., and M. E. Carolan. 1976. Daily instability of communal roosting in the Neotropical butterfly *Heliconius charitonius* (Lepidoptera: Nymphalidae). *J. Kans. Ent. Soc.* 49:346–59.

Species Index

Subject Index

Abdomen, structures of, 65, fig. 47, 66

Abdominal brushes, 55

Abdominal heart pumping, 45. *See also* Heart: muscular contraction of

Aberrations: autosomal recessive, 165 (*see also* Genes: expression of); and melanization, 165 (*see also* Melanic scales; Melanin); pigment fading as, 164–65

Accessory pulsatile organs, 45. *See also* Circulatory system

Aestival morphs. *See* Seasonal polyphenism: aestival morphs

Alleles, 110–11, 156–57; variation in, 159–61. *See also* Genes: composition of

Allelochemicals, types of, 181

Allylglucosinolate, 183–84

Alpine climate, characteristics of, 80–81. *See also* Arctic climate

Anal cerci, xi

Anal veins, 48, fig. 31, 49

Androconial scales, 55–56; in courtship, 152. *See also* Mating

Angelicin, 130

Ant-butterflies, 69

Antennae, 14, 36; cleaning of, 32

Antiaphrodisiac pheromones in *Heliconius*, 191. *See also* Mate rejection, communication of

Antifreeze compounds, 84

Anus, 67, fig. 49

Aorta, 45

Apostatic selection, coevolved herbivores and, 188. *See also* Coevolution

Apposition eye, 33, fig. 20

Arctic-alpine butterfly convergence, 80–87

Arctic climate, 80–81. *See also* Alpine climate, characteristics of

Arolium, 42

Aromatic compounds in defense. *See* Defense: aromatic compounds in

Artificial diets, 186

Atrium, 67, fig. 49

Automimics, 140. *See also* Batesian mimicry; Müllerian mimicry complexes

Autosomes, 164. *See also* Genetics: of butterflies

Balanced seasonal polyphenism, 154–55. *See also* Polymorphism; Seasonal polyphenism

Basal area, xii

Batesian mimicry, 107, 142–48. *See also* Automimics; Müllerian mimicry complexes; *color plate section*

Battledore scales, 54–55

Biennialism of arctic and alpine butterflies, 91. *See also* Voltinism

Bilateral gynandromorphs: characteristics of, 166; genetics of, 166–67

Biological species concept, 172

Bivoltinism, 90. *See also* Voltinism

Body basking, 76, 85–87. *See also* Thermoregulation: behavioral; *color plate section*

Body structures, 17

Brain hormone, 20–22

Broods, 90. *See also* Phenology; Voltinism